# Demotorized

James Ruppert

FORESIGHT
PUBLICATIONS

ISBN: 978-1-9160522-2-2

# DEDICATION

David Bellamy. The Gilets Jaunes, all Motorists and my sister, Marion. Also a Motorist.

## CONTENTS

# ACKNOWLEDGMENTS

Peter Edwardson who has been running his excellent speedlimit.org website and most heroically charted petrol prices for a generation which are fascinating and horrifying.

Roger Lawson and Brian Gregory at the Alliance for British Drivers. They both talked to me which is far more than the AA or RAC Foundation ever did. I am not a member but will probably join. They seem to actually stick up for the motorist and Brian had a scientific understanding of the climate which helped hugely.

Also, Transport for London, The British Parking Association and the Motorcycle Industry Association who all took the time to answer the questions at the back of the book.

The motoringpodcast.com

Dee who continues to give support and advice, including, 'are you really sure about this book?'

Olivia who is better at working on cars than I am and if this book does anything it might help postpone the banning of cars long enough for her to enjoy many more years of motoring.

# Introduction

I love cars, always have.

Cars, allow people to be free. They use them for work, to go on adventures and help others. Motors are a force for good which outweighs any downsides, yet owners have been relentlessly persecuted by the authorities despite being law abiding and conscientious. The motorist has therefore been a very easy target for taxes and the blame for everything from that new fangled climate emergency thing to the overall decline in the quality of life. I think that's unfair.

Demotorized is partly a giant whinge on behalf of the average motorist. I thought it was right to detail some of the gripes as much as possible from a British perspective. That is despite the Americanised Z in the title, which I believe looks more graphically interesting and makes for a more attention grabbing cover. That and a picture I took 27 years ago in North London which seems to symbolise what is happening to the Motorist.

I believe that all the Motorist wants to do is finish their journey, get to work, or visit their mum. Trouble is, there's an ultra low emission zone, or a new road toll to contend with, meanwhile some soft headed climate extinction protesters have glued themselves to the dual carriageway. Not only that, the Motorists' car is subject to a manufacturer recall because the exhaust emissions were tested on the wrong sort of monkey. By the end of Demotorized this will all make some sort of sense.

This is a book I had not planned to write. I know the world is waiting for me to knock out quite a few more books with Banger in the title. The trouble is I was invited on a BBC Radio 5 Live debate one Saturday evening with Stephen Nolan.

I was on air to talk about the parliamentary select committee proposal, which suggested the private car should ultimately be banned. Obviously I disagreed.

Opposing me was an environmentalist lady and Stephen Nolan. She admitted to having a car, which was awfully convenient, yet called me an idiot for simply disagreeing. The presenter said that children were convinced about climate change, so there. There was an audible intake of breath when I said I didn't believe in a 'climate emergency' and even picked apart, "An Inconvenient truth" for them.

Despite the entertainment I provided, I haven't been invited back.

Obviously I want to live in a world that is brighter, cleaner and more fragrant than before. Genuine technical and engineering advances are to be welcomed to make the air we breather cleaner and the world better. I wrote Bangernomics in 1993, which detailed how to run a car, not just on the cheap, but as an alternative to being part of the buy new merry go round. What could be more environmentally friendly than that? I'm also a lapsed member of the Battery Vehicle Society, back from when it wasn't at all fashionable.

The evidence that the average motorist has suffered from being overtaxed by government, mislead by cheating car manufacturers and bullied by pressure groups is overwhelming. I may have missed some issues and I apologise. Every single day there are new stories, new statistics and viewpoints. My excuse is that this is how the changing motoring world looked at the end of 2019 and how much the motorist had to deal with. At this current rate of attrition I'd say there is five years left to save the car. I might not be joking.

Politicians have caused most of the problems through their stupidity and lack of understanding how cars actually work. They chose to believe that diesel engines were cleaner because they emitted less C02 and still think that electric cars are zero emission.

Motorists have been under attack from all sides ever since they proved it was possible to make your own way in the world behind the wheel, or tiller of a self-propelled vehicle in 1803, it may have been 1802, but no one is entirely sure. If Toad of Toad Hall was still terrorising the countryside in his, toot – toot, motor, a very bad example of a motorist, he'd at least be making a fuss about all this car related crap. He would have joined the Gilets Jaunes.

Just in case you wondered, I wore my yellow jacket throughout the whole writing process. Since then I have never taken it off.

Just in case you wondered, I still love cars. Motorbikes and pushbikes.

I haven't gone for lots of graphs and pictures to confuse the issues, just some silly Victorian/Edwardian lithograph cartoon things to lighten the mood.

I very much appreciate you buying Demotorized and if you have any comments then please get in touch.

James Ruppert - February 2020

*One of those 'Self Driving' Vehicles*

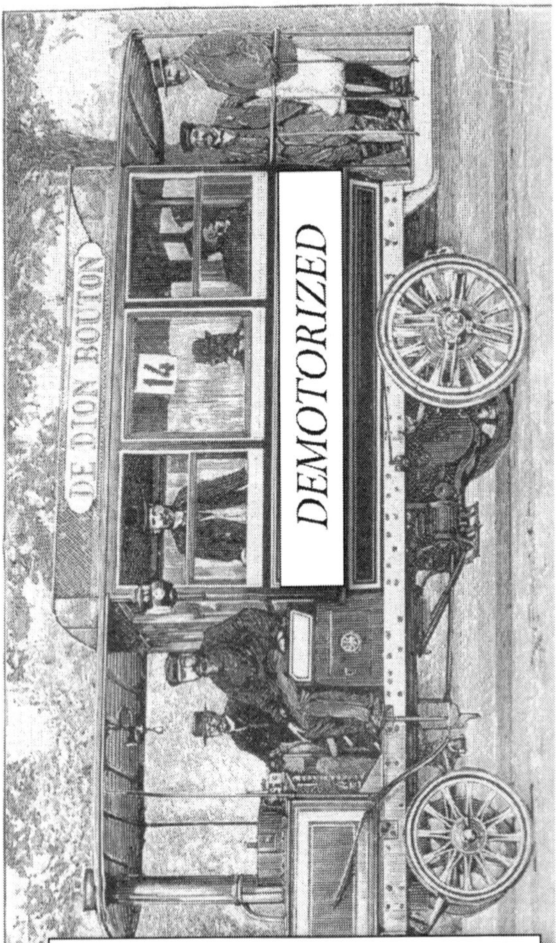


*The Demotorized Society on an outing in order to see if they can agree on anything...*

## **0. Demotorized Common Ground** - *Some things we can all agree on. Possibly.*

Although Demotorized is all about promoting unfettered motoring extremism and independence, fighting against authoritarianism and bureaucracy, sometimes you do need rules.

Learning how to drive properly was not a legal requirement for some time. That's because most drivers were the gentry who were born to do this sort of thing. Once the plebs got involved, standards were certainly needed.

The Road Traffic Act 1934 made the driving test a voluntary option, apparently to avoid the rush before it became compulsory in 1935. Interestingly, it was only for those who started driving after 1st April 1934. All the old fools driving around before then were in the clear, for the rest of their lives.

Even more interestingly anyone who took and passed the motorcycle test did not need to take another test to drive a car. That does make some sense, so much so we recommend it be reintroduced in a later chapter.

A constant, has been the best selling book, called 'The Highway Code', which came out in 1931 and cost 1d. You needed to read and learn most of it, so you could in the old days, answer the random questions of the examiner after the driving part of the test was over.

Everything became very 21st Century when in 2000 the touch-screen theory test was introduced on 4 January 2000. There was more button pushing when driving test candidates had to start following directions from a sat nav at the end of 2017. In June 2018, learner drivers were allowed to take motorway driving lessons and from November hazard perception part of the theory test started to include video clips.

That is a selective, very potted history of teaching Brits to drive. It is far from perfect, but on the whole the system seems to work, except when it doesn't. Car driving test cheats have been on the rise with increasing amounts of fraud in the exam itself and the last time that the Driver and Vehicle Standards Agency (DVSA) released any figures there were 158 impersonation cases in 2004. Presumably there have been an awful lot more occasions since where a 'lookey likey' took the test, for money, for someone else. These were just the ones who have been caught.

The best that we can hope for is that the person behind the wheel isn't under the influence of anything that could affect their driving ability. Many people reckon that in the old days you could drive drunk with no consequences at all, provided you didn't end up in a ditch or in the rear end of a police car. The 1872 Licensing Act created offence to be drunk while in charge of carriages, horses, cattle and steam engines. You could get fined and even get hard labour for that.

The 1925 Criminal Justice Act made things even clearer

when it became an offence to be found drunk in charge of ANY mechanically propelled vehicle on any highway or other public place and the driver could be disqualified. By the 1930 Road Traffic Act you only had to be in charge of a vehicle. That meant having a set of keys while sitting in the vehicle and it wasn't just drink, but drugs too. There were lots of subtle changes, but the really big one was the Road Safety Act of 1967 which introduced the first maximum legal blood alcohol (drink driving) limit in the UK. It was set at a maximum blood alcohol concentration of 80mg of alcohol per 100ml of blood or the equivalent 107 milligrams of alcohol per 100 millilitres of urine. Then there was the breathalyser.

It is at this point that drivers really started to get annoyed. They could be randomly brethalysed without reason, for instance, that they were not impaled on a tree trunk. Many viewed it as an attack on personal freedom, but rather more importantly some suggested that the rural pub would die as a result. It didn't help publicans, but going out for a drink and a drive, not something Mungo Jerry would ever endorse, never made any sense. Having a designated driver was always going to be the way forward. By 1991, the Road Traffic Act introduced 'Causing death by driving while under the influence of alcohol or drugs'. Then in 2014 in Scotland Reduced the Drink Drive Limit to 50mg of alcohol per 100ml of Blood.

Drugs have always been a difficult one. If there are

drugs in a drivers system, and this includes legal drugs, there is no need to prove that they are impaired in any way. However, Police can stop the driver if they suspect drugs use and can test at the roadside using either a drug screening device, or a "field impairment" test, you know, touch your nose, stand on one leg etc.

Never mind banning drink and drugs, what about phones? It is illegal to text, or answer calls unless a Bluetooth or other connected system is fitted. It is also not permitted to idly flick through music playlists, social media or messages on a smart phone.

However, it is possible to use a map app on a telephone, but certain rules must be followed to stay on the right side of the law. That means a correct cradle or holder is required to hold the phone when driving. Before driving, the destination should be keyed in, not in traffic, or at the lights, but only when parked.

The police can stop drivers if they believe they are distracted or not in control of the vehicle at any time. The phone screen must be positioned so that it is not regarded as either a distraction or an obstruction, according to the Highway Code. It states that both the windscreens and windows must be kept clean and free from obstructions to vision.

Right then, if the driver is qualified, sober and not distracted by their smartphone, what about the vehicle itself?

That would be the MOT test introduced in 1960. Originally a basic inspection covering brakes, lights and steering all carried out on the vehicle's tenth birthday. Also known as the "Ministry of Transport Test" most cars were only fit for the scrapheap after a decade. Not surprisingly, the age for testing was reduced to seven at the end of 1961, and then in April 1967 it was cut to three years. The list of items tested has steadily increased.

These are the basics we probably all agree on. Then again, you can argue about the details with all of them. There are shortcomings when it comes to the driving test and whether the legal the drink-driving limit should be zero nationally. When it comes to testing a car it certainly has become a lot more environmentally focused and the last thing we want is a smoky exhaust, or engine, because then there is probably something seriously wrong. An owner who does not care about their driving, or car, ought to be Demotorized.

For the rest of us, who just want to go about our daily, driving business, it would be great if we were simply left alone. Motorists, especially the law abiding ones, are just too tempting, lucrative and soft a target for governments, criminals and ne'er do wells who delight in telling others what to do.

*The Top Gear Team in 1885*

*A lazy Red Flag Bloke keeping his arse warm...*

## 1. Red Flag. No Bull - *An unnecessary history of carriages without horses.*

Before we get to the flag, it is worth looking at just how the world first became self-propelled. There might be hints as to why the authorities thought they should be controlled and the public protected from the dangerous contraptions that were lashed together and let loose on the King or Queen's Highway.

**The French Tank** Historically the very first car wasn't a car at all. In 1769 Nicolas-Joseph Cugnot actually constructed a great big steam powered tricycle tractor. It wasn't pretty, but it certainly was clever enough. With the boiler overhanging the front wheels the steering and brakes were not great, but no one was very bothered because this was essentially a tank, or more accurately a 'Boiler'. Military engineer and army captain Cugnot was not after a more convenient and less tiresome way to get to the shops, but a method of getting cannons to the front line.

His boss, Napoleon Bonaparte, was all about unifying Europe and this was one way of achieving that aim. The trouble was that Cugnot's two and a half tonne steam powered 'Mark 2' Tractor in 1770 wasn't great in the handling department. A bit like an unsophisticated Sports Utility Vehicle based on a Morris Marina. Losing control, the Cugnot, steered by Cugnot, took out a wall. The French, or rather Cugnot had invented a tank, but didn't realise it.

However, because this was a top secret vehicle the wall demolition was covered up and not mentioned until 1801. Also, not mentioned is that a more convincing and less destructive steam vehicle was built in China several years before, by a Jesuit priest. Father Verbeest was attached to the Court of Chien Lung, Emperor of China. In 1672 he apparently built a steam car, part of the process of seeking to convert the country to a Jesuit way of thinking and a Catholic way of worshipping. That didn't work, but apparently you could see it at the Winter Palace in what was Peking, certainly into the 1940s. Presumably, the new People's Republic of China were not interested in Catholic cars, so like a lot of unfortunate Chinese citizens, it disappeared.

Cugnot's tricycle tank tractor was parked in the Conservatoire National des Arts et Métiers in 1800, so the French win the 'who made the first sort of car' award. In Blighty though the great steam pioneer James Watt had planned to build a steam road vehicle, but was already very busy in the stationary engine business. Watt did take out a patent on a road going steam engine with partner Matthew Boulton, whist employee William Murdoch made steam powered models. Indeed, his unremote and not very controlled, Steam 'Tamiya' even scared the Vicar of Redruth into believing he'd seen the devil.

**British Steam Cars** The real significance of Murdoch's steam engine was a connecting rod from a beam to the axle. This made it the first wheeled vehicle to be powered by reciprocating action, a bit like a real car. Anyway Murdoch was busy with so many things for Watt and Boulton and forging the industrial revolution that any plans to make cars were put on the back burner. Dig a bit deeper and you discover that Watt was something of a weird fish. He was deeply jealous of the clever Murdoch and reliant on Boulton's cash injection to get their business off the ground. So he shelved any plans at all to make a steam car, but the patents for them stopped anyone else exploiting a gap in the personal transportation market.

Another steam pioneer though did see some potential in having private transportation. Watt's major rival was building lighter, high pressure and more efficient steam engines that leant themselves to being a lot more portable. Richard Trevithick made what is the world's first self-propelled passenger-carrying vehicle. Basically, a car. He built a couple of them and the most notable was the London Steam Carriage, mainly because that's where he used it. Some sources say it was 1803, other publications 1802, which I'll go with. It was no less preposterous in scale than a modern day SUV as it weighed in at 1.9 tonnes, with a fancy carriage perched on top of a couple of large carriage wheels and a small one to steer. Essentially this was less of an SUV and much more of a proto Reliant Regal van, which seated up to eight. Trevithick drove it in and around London. The perfect attention seeking leviathan today, but back then it was a puffing, steaming and rather expensive novelty.

The private Steam Utility Vehicle was way ahead of its time. It even had its crank powering the road wheels through a series of interchangeable gears, which was a real glimpse into Britain's bright engineering future.

According to contemporary observers it travelled "faster than a man could walk" which was good, but clearly not faster than a trotting pony. Contemporary estimates say 8 mph, which is twice as fast as a brisk walker. A gentleman called Captain Vivian having braved one ride and being offered another declared that he was more likely, "to suffer a shipwreck on this steam carriage than on board my own sailing vessel." There's someone you wouldn't want to be out with in a Force 10, he sounded like a wuss to me.

There didn't seem to be much in it for those who could actually afford a London Steam Carriage. It cost £200 to build the carriage alone and no one knows how much the engine was, so the expenditure was mainly Trevithick's time and lots of cast iron. At the time the social issue was, why would a member of the landed gentry want to drive themselves? Although some young man who was in the Household Calvary might have seen the rakish potential of gadding about in a high performance steamer, rich old farts couldn't be bothered. Gentlemen hired coachmen and the support network that came with horses and carriages. So there were zero sales. Presumably with some better public relations, marketing and structured dealer network he might have done really well. Instead, The London Steam

Carriage was broken up for parts, the main part being the engine, which wound up making the hoops for beer barrels, so something useful came of it.

Trevithick, never had much luck with any of his brilliant innovations and inventions. It is worth reading the biographies if you feel that your own genius and hard work has not been rewarded or appreciated. He at least had an interesting yet unremittingly tough life, travelled extensively, but died a pauper at 62.

Meanwhile, steam propelled mass transportation was catching on. Putting steam engines on tracks so they could go in circles or on defined journeys seemed to be the profitable way forward. There were though proper engineering mavericks determined to head in their own direction. This depended on proper roads being built by Thomas Telford and John Macadam.

**Bendy Bus** It took a while to get here, but now we really are motoring. Or steaming and heading towards the self-interest groups that would resist road based progress. Say hello to Sir Goldsworthy Gurney who was an English surgeon, chemist, lecturer, consultant, architect, builder and all round top hatted good egg who fell head over heals in love with the whole concept of mechanical transport. One report I read said that he became 'violently' enthusiastic about it all. Good man.

He originally set up shop in Oxford Street in 1825 to work on a steam carriage. His first patent described "An apparatus for propelling carriages on common roads or railways - without the aid of horses, with sufficient speed for the carriage of passengers and goods". He developed something described as a 'blastpipe', which used steam to increase the flow of air through a steam engine's chimney. That increased the draw of air over the fire boosting the power to weight ratio of a steam engine, making it quick and efficient.

This was an expensive business so money was either tight, or always running out. The first extensive trial in 1829 took him from London to Bath and back. Parts broke, but Gurney reckoned the average was 12mph and presumably when going downhill with the wind behind his carriage, touched 20 to 30mph. If true, that's faster than a speeding horse and presumably it wouldn't run out of puff so quickly, provided the boiler was kept stoked.

Gurney wasn't the only one in the steam carriage business, but he had more of a talent for getting attention. As early as 1833 Walter Handcock had designed some carriages, which ran a passenger service between Oxford Street, Edgware, Regents Circus and Uxbridge. Yes he was operating it just like a bus. Trouble was not everyone was happy to sit on top of these contraptions, so the commercial case was not very strong. Gurney even separated passengers from the potentially a boiler shaped bomb by putting them on a trailer. That's right, Gurney had invented the articulated bus. The Bendy Bus.

Despite improvements, the state of Britain's roads was

pretty much like they are now, a potholed mess. Pneumatic tyres had yet to be invented and there was also considerable opposition, or NIMBYS (not in my back yardists) a.k.a. moaning minnies. Curiously there were few reports of injuries or serious road traffic incidents involving these gallumping, steamy vehicles. The Times Newspaper had a recollection from the daughter of an operator several decades after the incident date relating to a journey from London to Bath in July 1829. "I never heard of any accident or injury to anyone with it, except in the fray at Melksham, on the noted journey to Bath, when the fair people set upon it, burnt their fingers, threw stones, and wounded poor Martyn the stoker".

The public could be rough and ready and react badly to new technology, especially one that could deprive them of a living and take it out on innocents like Martyn. But sticks and stones from the great unwashed public was nothing once the established order had their posh noses put out of joint. Not surprisingly, the wealthy horse-coach proprietors, got together with the landed gentry and magistrates of the district. They were either in Parliament or had good friends in the House. As a result they managed to pass local turnpike bills putting tolls on steam carriages. £2 was levied on each steam carriage journey, whilst the toll for a horse drawn carriage was just 2 shillings.

Not only that, the squires and magistrates would get together and cover roads with a layer of loose gravel, at least a foot deep. Effectively putting physical barriers in the steam carriage's way. They also spun negative press coverage when there was any incident or controversy. One of Gurney's articulated jobs had a problem in Glasgow and the social media of the day, newspapers, ripped him to shreds to such an extent that he was bankrupted, for a frankly staggering £232,000.

**Red Flag & Regs** A House of Commons Select committee from 1831 to 1835 looked into Gurney's Case. They noted that he ran one of the most successful operations of its kind and even took out patents to cover himself. He managed to negotiate contracts that meant he would end up carrying passengers for lower fares than horse carriages.

That select Committee concluded that steam carriages were safe, quick, cheap, and less damaging to roads than horse carriages. As a result they were of benefit to the public and the tolls that discriminated against steam carriages should be dropped. Incredibly, a bill was drawn up to this effect and we could have entered a golden era of road travel. This bill effect was even passed by the House of Commons but surprise, surprise, was turned down by the self interested House of Lords. The road locomotive was dead.

Just to make sure that it was at least six feet under there was something known as the 'Red Flag Act'. Officially, The Locomotive Act 1865 was intended to regulate the use of Locomotives on Turnpike and other roads for 'Agricultural and other purposes' which was a great big catch all. Although the impression given is that Parliament specifically passed this vindictive act to get those new fangled vehicles off the road, the truth is that the petrol or electric powered private car had not been invented yet. It seems there was an intention to keep a cap on large steam powered farm machines with massive steel rimmed wheels.

Just for entertainment value then let's see what this act was supposed to cover. Well first off these locomotives weighed up to 14 tons, and were up to 9 feet wide. Such self-propelled vehicles should have a crew of three. However, if the vehicle was attached to two or more carriages an additional bloke was to accompany them and another with a red flag was to walk at least 60 yards ahead of each vehicle. He also had to help horses and carriages get out of the way.

The vehicles were also required to have functional lights, and must not make a nuisance of themselves by sounding whistles or blowing off steam whilst on the road. There was an overall speed limit of 4 mph, but that was slashed to 2 mph in towns. Break any of these rules and there was a substantial fine of £10.

Local authorities still had the power to ban whatever sort of vehicle they wanted all together. By 1878 the law was softened a bit as there was a requirement that each locomotive had to be preceded by a bloke, but there was no longer any mention of a red flag. It was though a formidable symbol and shorthand for the dull hand of authoritarian suppression. In effect, spoil sports.

**European Carriages without Horses** Things were stirring in Europe at the time, or more to the point, horseless carriages, rather than those puffing locomotives. Karl Benz was the first to recognise that there was a business in replacing a horse with an engine and set up a production facility to prove it. The 'Motor Wagen' was his big idea and he rolled out the first car, a three-wheeler in 1886. Meanwhile Gottlieb Daimler had been working intensely on internal combustion engines producing the first unit to run on petrol in 1883 followed two years later when Daimler bravely went with one wheel less than Benz's car. Yes here was the world's first motorcycle and the very last time that Daimler ever built a bike.

Clearly developments were afoot across the channel whilst British engineers felt restricted by the Locomotive Acts. Frederick Simms made a difference, born of an English father and German/Italian mother in Hamburg, he had made the acquaintance of Daimler in Germany. Simms acquired the rights for Daimler's patents in Britain and the Empire, but excluding Canada in 1890. However, as making cars seemed to be a bit of problem, with or without a red flag, he decided to use the Daimler engines on boats, although the posh word would be launches back then. In the meantime securing refined fuel for them meant he inadvertently created the word petrol. In anticipation of the law changing, or perhaps because he had a spectacularly good tip off, formed the Daimler Motor Syndicate in 1893.

**Her Majesty's Autocar** Momentum was certainly building during 1895, which seems to have been the turning point. Automotive enthusiast, Sir David Salomons put on the first motor show in Britain in the glamorous location of Tunbridge Wells in Kent. Simms was there promising that he could supply cars to all and sundry. Solomons also formed the 'Self-Propelled Traffic Association' so that he could campaign against the so-called Red Flag Act. We will come back to that at some much later date. Autocar were there to report on it, but they had yet launch themselves as the world's first car magazine which hit, though not in an uncontrolled way, the newsstands a couple of weeks later.

The Autocar was founded by William Iliffe and Henry Sturmey. Iliffe's family owned a printing company while Sturmey was editor for The Cyclist magazine. They saw the potential when a horseless carriage came to their offices and Sturmey was right behind it. The first twelve-page issue proclaimed that it would cover 'the interests of the mechanically-propelled road carriage.' The Autocar tripped off the tongue rather more easily. Sturmey went on to explain what it actually was, a horseless carriage, automobile carriage, automatic carriage: autocar. All these names have been used to designate the latest production of the ingenuity of man, the motor-driven road carriage, irrespective of whether steam, electricity, hot air, or petroleum be the motive power." Those are prescient words, which will repeatedly filter through Demotorized pages.

All this motoring activity was building up to one thing. Confusion. Or at least some very complicated business arrangements. The British Motor Syndicate was starting to register Daimlers with Harry Lawson at the helm. He confirmed that the company would be floated and then buy Coventry Cotton Spinning and Weaving Company, using their premises and renaming it Motor Mills. Meanwhile he was busy setting up the Great Horseless Carriage Company to get more finance for Daimler and unify loads of patents to get a proper monopoly going.

This was all building up to The Locomotives and Highways Act of 1896. Now, if you had a 'light' Locomotive, under four tonnes, unladen, it now needed lights, a bell and they could tear along Queen's Highway at up to 14 mph. There never was a requirement for a bloke to be present with a flag. However, that didn't stop them having fun on the Emancipation Run from London to Brighton to celebrate the passing of the Act. A red flag was produced which Lord Winchelsea promptly set light to before they headed to the coast.

Autocar was there reporting on all this flag burning and appropriately printed their November 14th issue in what might have been called generations later by a British Leyland dealer, Damask Red. On the day Harry Lawson said, "Motors would not become tired nor fall down nor bolt." Even so, not all the cars made it to Brighton, out of 33 that left after breakfast at the Charing Cross Hotel, just 17 got to the Metropole Hotel.

Never mind, a point had been made and they were at the seaside with access to sticks of rock and slot machines.

1896 was year zero for motors in Britain. Horses were heading for retirement/glue factory, businessmen had some excellent money making opportunities, government fresh taxation scenarios, whilst all the driver wanted to do, was go motoring.

The self-propelling vehicle was here to stay.

The authorities were just starting to limber up.

*One of the Seven Sisters wearing her favourite oil Derrick hat...*

## 2. Greasing the Wheels - *Inside the Oil Business.*

Before we get to the history of taxing the actual means of movement it is worth following where the money actually goes and what a gallon costs before the government takes their cut. Lots of oil companies are more than happy to spell this out so we don't actually hate them so much.

The simple fact is that the wholesale price of crude accounts for nearly one-third of the price of a litre of petrol or diesel. The rest goes towards taxes, the costs of running service stations and the profits, which retailers need in order to stay in business. It is important to remember that all these prices are very volatile on a worldwide basis. Plus there are always differing supply and demand levels. Refining margins are around 3p a litre, while retail costs are about 7p.

That of course leads to lots of moans when it comes to any crude barrel price cuts being passed onto the motorist. Usually they are not.

It is worth bearing in mind that diesel and petrol are independently traded. The price difference will vary and it can be seasonal when for example, more diesel being is used in winter, so the price rises.

According to the oil companies the forecourt price will always track the wholesale price trend and ultimately the crude price. Because tax is around two thirds of the pump price any drop in the crude price will only have a small effect on the pump price, so it seems as though they don't bother passing on any savings.

That's odd though, because prices seem to vary between service stations around the country by the odd few pennies. That is down to transporting the fuel to forecourts and running the service stations themselves. The oil companies will only control the sites they own the rest is up to the independent operators who quite rightly set their own prices when it comes to a packet of crisps or a barbecue kit. Obviously a station in the back end of nowhere is going to cost more to run than a busy one in a built up area, hence the variations. Profits are thin, selling fuel generates about 2.7p a litre of profit. That explains why their fizzy pop and sausage rolls can be so expensive. At least service stations are convenient. Open when every other shop is shut, plus you can fuel your car properly and body marginally.

Shell, for instance, claim that we have some of the lowest fuel prices in Europe, before tax that is. The two thirds of tax is fuel duty and Value Added Tax (VAT). At the time of Demotorized being published, fuel duty in the UK is 57.95 pence per litre, regardless of petrol and diesel prices and VAT is 20% Value Added Tax.

Here's the problem then. If the base cost of petrol falls, so the proportional impact of the fixed tax on the overall price actually increases. The mathematical result of this is that when oil prices rise, the incremental effect is felt quickly and in full. When oil prices fall, even if retailers were to pass on the decrease straight away, the benefit will be less because of the fixed element of fuel duty.

**Sister Act** Who though really sets the price of fuel? Who controls it? Initially that would be the Seven Sisters. Anthony Sampson wrote a brilliant award-winning book, of that name in 1975. The original seven companies who worked together were Anglo-Iranian Oil Company, which is now better known as BP. Gulf Oil, which like several others became part of Chevron, Royal Dutch Shell, the Standard Oil Company, another company absorbed into Chevron. The Standard Oil Company of California and the Standard Oil Company of New Jersey, which became (Esso, later Exxon, now ExxonMobil) and yet another regional Standard Oil Company, this time of New York and now part of ExxonMobil). Finally, there was Texaco.

Enrico Mattei coined that phrase "the Seven Sisters" to describe these monopoly companies controlling the world's oil supply in the Middle East. He wasn't happy. A clever chemical businessman and Italian resistance fighter, he was charged with sorting out and shutting down the state owned Agip Oil Company after the Second World War, but instead expanded it. That was despite the Seven Sisters and his death in a company jet in 1962, which is the subject of many complicated conspiracy theories. He probably had a point then about the way they operated and paid with his life.

Certainly in the wake of the First World War, every government realised how vital oil was. Just not to fuel, cars, planes and tanks, but also the supply lines. During the conflict Germany had struggled to refuel and keep their armies busy. Meanwhile Britain had access to Persian oil from BP, plus Mexican and East Indies oil from Shell. The bulk of the oil though came from the United States of America and a quarter through Exxon alone.

Post war it was all about securing oil supplies as America struggled to satisfy their huge demand. The Royal Navy switched fuel from coal to oil and the motor industry was growing at a fantastic rate chugging along like a Model T Ford because absolutely everyone was driving one. What the oil companies discovered was that competition between themselves actually caused uncertainty. Not for the general public that used the bi-products, but oil companies. They believed that close co-operation and collusion would help make the world work the way they wanted it to.

It all dates back to a meeting of three men in August 1928, in the Scottish highlands at Achnacarry Castle. A Dutchman Henry Deterding who founded Royal Dutch Shell, American Walter C. Teagle representing the Standard Oil Company and founded by John D. Rockefeller, plus an Englishman, Sir John Cadman, the director of the Anglo-Persian Oil Company, the future British Petroleum.

What they decided to do was set up production zones, share transport, agree prices and effectively dominate the world, by controlling its oil. That became a reality when the other four companies joined them. They achieved this by installing dictators, indulged in massive government level corruption and even helped kick off wars. The oil market was a false level playing field and ultimately the consumer paid a higher price. There are book loads of great and sometimes terrifying stories about the world's largest and most unchecked cartel. At the end of the 1960s, the Seven Sisters, effectively the major oil companies, controlled 85 percent of the world's oil reserves. Then came another cartel. An even less benevolent one motivated by politics, which pushed prices up and supply down to create a whole new age of uncertainty. Say hello to OPEC.

Before we get into all that, the British public did not have to wait very long for rationing to come back into their lives courtesy of the Suez crisis and specifically for petrol. Here's the simplest and briefest potted account I can manage. On October 29, 1956, Israeli armed forces pushed into Egypt toward the Suez Canal after Egyptian president Gamal Abdel Nasser nationalized the canal in July of that same year. The Israelis were soon joined by French and British forces. In the end, they withdrew. In the meantime there was rationing as supplies of fuel from the Middle East were blocked resulting in fuel shortages. Before the rationing there was a 'temporary' tax increase which put petrol up by 1s 5d a gallon.

Drivers didn't have to wait very long for an even more organised approach to making sure they paid even more for a gallon. In 1960 the Organization of the

Petroleum Exporting Countries (OPEC) was created in Baghdad with Iran, Iraq, Kuwait, Saudi Arabia and Venezuela as founder members. There have been a lot of country sized comings and goings ever since and we won't bore you with those. However, their intention has always been the same, "to co-ordinate and unify petroleum policies among Member Countries, in order to secure fair and stable prices for petroleum producers; an efficient, economic and regular supply of petroleum to consuming nations; and a fair return on capital to those investing in the industry."

Nice intentions perhaps and doing the best for your country is what countries should always do. Trouble is all this activity came with an extra, unwanted ingredient of fairly nasty politics. They were not behaving like the 'Sister' gangsters anymore, these were full on Bros' with automatically weaponised views on the world. It wasn't just about money, they were dedicated to causing trouble on a worldwide basis.

1973 is when the world's relationship with fuel changed forever. Basically it became bloody expensive and the globe a much more dangerous place. This began with the Yom Kippur war between Israel and a number of states in the Middle East, which started on 6th October. The Arab members of OPEC implemented an oil embargo on 17 October, initially applying to the United States but quickly expanded to Western Europe and Japan because of their perceived support, which those countries were giving to Israel. The embargo was accompanied by significant increases in the price of oil, which went up from around $3 per barrel before the war to over $11 per barrel by January 1974 and around $15 per barrel by the end of that year.

**Welcome to the Dark Oily Ages** There seemed to be a pattern emerging. Mainly, don't source your fuel from an unstable region of the world. And so it came to pass. As the Shah of Iran attempted to modernise his country he ran into religious opposition. There was some violence and the cheerful, moderate, man of the people, Ayatollah Ruhollah Khomeini returned from French exile once the Revolutionary Guards had ousted the Shah in July 1979. The Ayatollah installed a militant Islamic government and suddenly exporting oil became less of a priority. Especially when you are busy returning your country to the dark ages.

At this point the price of crude oil doubled from around $12 to $26 a barrel. In Britain the unthinkable happened, the £1 gallon had arrived. It didn't end there as the price kept on rocketing upwards and finished at £1.20 a gallon. Ever wondered why petrol was sold in litres rather than gallons? Presumably it was when those clever people in the marketing department (you need to market petrol?) decided the price would look better in litres and the price of a gallon would be hidden. There is certainly some truth in that.

Actually, it was metrification, which had been with the British people since 1971 and this was all to do with having insufficient digits on the pump display. Some machines could be converted, but certainly by the early 1980s most petrol, except at the most rural outpost, was sold by the litre. There remains the conundrum that fuel consumption is still measured in gallons. That's because the range of figures from the least to most economical cars is all in double figures and can be stated in whole numbers while still seeing subtle variance between similar cars. Stating economy in single digits and fractions like in the L/100km makes it seem like there's not much difference between very dissimilar cars

**North Sea Spoiled** At this point it might be worth pausing to think about the fact that Britain probably didn't need to worry too much about oil imports. There was something under the North Sea. Indeed, British Petroleum (BP) discovered the giant Forties oilfield in 1970 and seven years later it was producing 500,000 barrels of oil a day. It made all sorts of financial and economic sense to drill for oil in the dangerous and rough North Sea when the first Middle East oil crisis had poked prices up to around $16 a barrel.

Britain as a nation got little from the North Sea until the mid-70s when the fields were established. By the mid-80s, revenues from both oil and gas were worth more than 3% of national income. Some say that should have been invested in low risk stocks like Norway did with their oil income. Experts at PricewaterhouseCoopers estimated that the fund would have been conservatively worth around £450bn by 2008. Instead that money went on tax cuts. Nigel Lawson slashed income and other direct taxes, as top rate of tax came down from 60p in the pound to just 40p by 1988 and he also reduced the basic rate of income tax to 25%, but there was always a prevailing rate of 15% VAT to keep everything else expensive.

With that spare cash people just went and bought houses. Meanwhile the national assets like British Gas were privatised. There is a bad spin to put on this, namely that the handing over of what are in effect the crown jewels at knock-down prices to corporations and supposed investors, who then skim off the profits, is not the brightest thing to do. Except that if the previously nationalised industries had been beacons of efficiency and real national assets, they may have a point. For instance, British Leyland wasn't a great example of how to run a company. Perhaps the income from oil should have been set aside but the truth is that governments don't actually have their own money, or operate with a long-term view. It is all about spending what isn't actually their money, now. Norway thought differently.

In 1974, they decided that oil wealth should be used to develop a "qualitatively better society". You can argue what that is, but the oil industry including the actual fields are publically owned and the fund adds up to more than $1 trillion in assets. It effectively made Norway's modest population worth $195,000 each. Then again, if you are from Scotland you might also have a point that the revenues should at least have been funnelled through their country so that the population might be better off. Or not. Governments are like that. Inconsistent.

It would be nice to think that Britain, which was a net exporter of oil at one point, could have slashed fuel costs to pennies a gallon. That was never going to happen with an economy as old as the British one and as fundamentally corrupt. The bureaucrats and politicians were also addicted to ever more creative ways to tax the driver.

Then there is the curious case of 'Peak Oil'. The online definition is "Peak oil is the theorized point in time when the maximum rate of extraction of petroleum is reached, after which it is expected to enter terminal decline. As of 2019 peak oil forecasts range from the early 2020s to the 2040s, depending on economics and how governments respond to global warming."

Everyone, or at least the media, used to be obsessed with the subject. If there was not a big story about an imminent ice age, then there were books and magazine features telling us that oil was running out. Time magazine even had a famous cover with a fuel gauge registering empty. Never mind the climate excuse, and fossil fuel bad explanation which were invented later on, there was an obsession with falling oil reserves. Until it was completely forgotten about.

What's happened is that the original definition as in 'the oil is running out' has now been changed to mean 'when there is peak oil demand', but that is a slight digression. The origins of the phrase can be traced back to 1956, when geologist M. King Hubbert at Shell Oil Company predicted that oil production in the lower U.S. states would peak sometime around 1970 and then tail off in some sort of bell curve chart. He was proved right.

He believed that the principle could not be applied on a worldwide basis. So when Hubbert turned his sights to global oil production in 1974 he predicted that the world's peak oil production would occur in 1995, assuming that the then current production and use trends continued.

President Jimmy Carter made a speech in 1977 about the energy crisis, "...just to stay even we need the production of a new Texas every year, an Alaskan North Slope every nine months, or a new Saudi Arabia every three years. Obviously, this cannot continue."

Slightly more recently a bloke from Greenpeace called Rex Weyler said, "Oil company cheerleaders proclaiming huge supplies of oil are dead wrong. Peak oil is as real as rain, and it is here now. Not 2050. Not 2020. Now." His now, was 2012.

Clearly predicting the future is not a precise science. Despite rises and falls in OPEC production, the Gulf Wars and the constant price manipulation, oil never seemed to go away. It just seems to cost a bit more. Those predicting its disappearance had to keep on upping their estimates of what could be recovered. The main explanation for all this abundance is the seemingly exponential availability of shale oil.

Hydraulic fracturing, better known as fracking has opened up numerous oil fields in areas that was once considered empty or too costly to develop.

Oil industry analysts often describe oil resources in terms of conventional and unconventional oil. Conventional oil describes oil that's available through more traditional, less expensive technologies like the oil wells in Saudi Arabia.

Saudi Arabia's national oil company, Saudi Aramco, pumped up production to 11.5 million barrels per day (mbpd) in the early 2010s from about 9.5 million in early 2009. Increasing their production to the highest amount for decades, along with the rest of OPEC. They overtook Russia who had held the top spot for several years. In third was America on 9mbpd. However, the latest figures (2019) show a big reversal.

The United States is the top oil-producer, with an average of 14.86 million mbpd, which accounts for 15.3% of the world's production. Down from 15.12 million mbpd in 2015, but it was enough to land the United States in the top spot, which it has held for the past four years. The United States overtook Russia in 2012 to be second, and then Saudi Arabia in 2013, to become the world's top oil producer. Much of the increased U.S. production is due to fracking in the shale formations in Texas and North Dakota.

According to the BP Statistical Review of World Energy for 2018 they seem to admit that oil isn't running out anymore by using some very fussy language to reinterpret what the phrase means. "Peak oil demand signals a break from a past dominated by concerns about adequacy of supply. A shift in paradigm: from an age of scarcity (or, more accurately, 'perceived' scarcity) to an age of abundance, with potentially profound implications for global oil markets as they become increasingly competitive, and for major oil producing countries as they reform and adjust their economies for an age in which they can no longer rely on oil revenues for the indefinite future."

It is worth swerving back towards Norway. Just when it looked as though their oil revenues were on the decline a massive new oil field called, Johan Sverdrup was discovered. In 2019 it was already producing 350,000 barrels each day. When it hits its peak in late 2022, the field is expected to produce almost double that, or 660,000 barrels per day.

"The person who will switch off the lights on the Norwegian continental shelf is not born yet," Prime Minister Erna Solberg said in late 2018.

There is no such thing as peak oil.

We have clearly been lied to or at least been fed bad science by our world leaders, environmentalists and experts. There seems to be plenty of oil around to be burnt in cars. At the very least Saudi Arabia should be empty by now and clearly it isn't.

Very finally, is oil really composed of dead dinosaurs? Seems a bit far fetched doesn't it? Has anyone ever done a DNA test on a pint of unleaded? Well that's for another book.

*Vanasaurus Rex*

*Tax Office officials randomly picking off motorists in their spare time.*

## 3. These Fuellish Things - *Taxing what moves us.*

It didn't take long for the British Government to realise that as well as turning fuel into forward movement for the motorist, it was also going to be a serious revenue stream.

Giving the motorcar just over a decade to gestate and grow in popularity and really get started meant that the time was right to start hitting it with some serious tax.

Petrol duty was introduced in 1909 at 3d per gallon, thanks to the Finance Act 1908. It was never going to stay at those levels and with the commencement of the Great War 1915 the exchequer was always going to need more money. So that reasonable rate was doubled to 6d. However, if you ran a commercial vehicle of some sort then your liability to fuel tax went back down to 3d because you were doing something useful.

With the First World War over, the Finance Act 1919 abolished the loathed fuel tax. Hooray. Trouble is, this was immediately replaced by the brand spanking new tax disc.

It was hardly surprising that the price of fuel dropped dramatically during the 1920s. As a result the government decided it was time to start squeezing the car owner for petrol duty again. So it was reintroduced at 4d per gallon in 1928 and the car tax stayed. The Government wasn't going to be that stupid again.

The Second World War provided the next opportunity for the Government to interfere with the allocation of fuel. Not surprising really considering that the safety, security and future of the country was at stake. Better to pump fuel into a Spitfire or Hurricane as the 'phoney war' turned into Dunkirk and then the do or die, Battle of Britain.

War was declared on 3rd September 1939 and the Government was ready by the 8th of September with its coupons for rationing petrol, which was called 'motor spirit' back then. These coupons could be collected from a Post Office. Applicants for the coupons were required to produce their car or motorcycle registration book, and they received two ration books one marked 'first month' and the other 'second month'. The books contained coupons for the quantity of 'motor spirit' allowed them according to the rating of the vehicle as shown in the registration book.

Initially the Government wanted to allow around 200 miles a month of travel. That meant everyone got a basic ration, but there were exceptions for those actively involved in the war effort. Specifically that meant Home Guard and Women's Land Army officers who received extra coupons. The rationing was also based on the car's horse power rating. Those with a 7 horsepower car, there weren't many, old cyclecars and BSA three wheelers, were allowed 4 gallons. 8 and 9 horsepower cars, like a Ford Model Y, got 5 gallons. For 20 horsepower and above, like a fancy Humber Super

Snipe, it was 10 gallons.

Each coupon represented one unit. At the time, a unit represented one gallon, although that could be changed. Cleverly the ration books were only valid during the period for which they were issued so they couldn't be hoarded.

Even if it was still possible, in July 1942 the government effectively banned driving for the hell of it. The official statement said, "As from 1 July the basic ration will cease to exist...These measures are designed to end pleasure motoring...The government wants all unessential cars taken off the road."

If you didn't have a good excuse to use your car then you were not going anywhere, unless you could afford to pay three times the per gallon price on the black market. If you are old enough to remember the BBC TV comedy, Dad's Army, then you'd be dealing with the pencil mustached, slick haired, private Walker as the spiv, who could get anything, at a price.

It was possible to tell legal from illegal petrol because it was dyed red. Not only could the police or army dip your tank to check for colour they could ask you to justify the journey, but also prove that you were taking the shortest possible route. Something of a challenge when all the road signs had been taken down to slow down any fragments of the German army who made it past the Home Guard at Warmington-on-Sea.

Not surprisingly, petrol prices went up throughout the war. From around 1s 6d at the outbreak of war, then in March 1940 to 1s 11d and by 1942 to over 2s. With each rise there was an actual cut in the ration.

Car ownership was far from commonplace at the time and was a luxury item for the few who could afford it. There were around 2 million cars on the roads in 1939. A year later in 1940 that had dropped to 1.4 million and by January 1944 was down to 700,000. The rest had been scrapped or were perched on bricks sometimes in garages, but also in the street. Car production was zero anyway as everything went into the war effort.

The basic petrol ration of five gallons was back on 1st June 1945. It was withdrawn again in summer 1947 and came back a year later. The small members, in number, not height, of the motoring public, were getting rather miffed. Rather than tolerate drivers turning to the recently demobbed (from Home Guard Private Walker, the government sought to control petrol with 'The Motor Spirit Regulation', which became known as the Red Petrol Act. Commercial road users got to fill up with petrol that was dyed red and everyone else had to make do with fuel that had a yellow tinge, if they could find any. Anyone who broke the coloured petrol law faced prison, a £1000 fine and a ban from driving. All that was pretty much the end of anyone's life, or their business if they sold the wrong colour fuel to the wrong person.

In 1948 an AA survey revealed that traffic levels were

less than half of what they were a decade earlier. That year also marked the very first Earls Court Motor Show and although the interest levels were sky high, there were very virtually no new cars to buy. Car companies were encouraged to export or die. Britain was bankrupt and had to manage repayments and reconstruction of the country.

Meanwhile, that country which was defeated by the sacrifice of the people and motorists of Britain, Germany, managed to end rationing in January 1950. Brits had to wait until the end of May. However, food rationing remained until July 1954.

The long suffering British public did not have to wait very long for rationing to come back into their lives courtesy of the Suez crisis and specifically for petrol. Before the rationing there was a 'temporary' tax increase which put petrol up by 1s 5d a gallon. Aubrey Jones, Minister of Fuel and Power, told journalists that businesses will be allowed an extra 100 miles a month in petrol, in addition to the 200 miles a month issued to normal car users. Farmers, religious ministers, and essential local authority workers will be allowed 600 miles a month. A third group, including doctors, midwives, disabled drivers and veterinary surgeons were allowed whatever petrol they needed.

The ration books issued from December 17th contained coupons to cover a four-month period. It all depended on the size of the car owned. So an 1100cc car got 6 gallons a month and a 2400cc car got 10 1/2 gallons a month. The temporary tax increase proved to be just that, but it still took a while to remove. Government statistics claimed that petrol consumption was reduced by 30% because of the rationing.

The only real upside of all this was it led directly to the design and production of the mini to counter the import of so-called bubble cars, mainly from Germany. Things then rattled along until the Yom Kippur war between Israel and a number of states in the Middle East, which started on 6th October 1973 and I think we explained in the previous chapter as concisely as we could.

Anyway, OPEC may have restricted oil supply as a political response to that war and their anti Israel stance, it also formed an economic strategy which boiled down to keeping oil in the ground meaning higher prices. In Britain motorists were asked to economize in the use of fuel. Peter Walker, the Secretary of State for Trade and Industry and cabinet minister responsible for the energy sector, asked on 24th October 1973 "motorists should endeavor to cut down on petrol consumption and to use public transport to a greater extent where that is available."

By 19th November, the government introduced a 10% reduction in all oil supplies and appealed for voluntary avoidance of driving at weekends and a voluntary reduction in maximum speeds to 50mph. This resulted in queues at the pumps and ultimately, ration books were distributed from 29th November 1973. More than 18 million petrol ration books were printed, 12 million "supplementary coupons", 20 million forms and 7 million envelopes which were sent to post offices and motorists to beat the Christmas rush. However, rationing never happened but the ration books stayed in drawers in case they were needed. A relieved nation were allowed to use them as kindling from July 1975 when rationing was off the agenda.

The rationing threat came back again at the end of the '70s, with the temporary closure of 200 petrol stations when Iran became volatile and went from a monarchy to a theocracy. The price of crude oil doubled from around $12 to $26 a barrel. In Britain the unthinkable happened, the £1 gallon had arrived. It didn't end there as the price kept on rocketing upwards and finished at £1.20 a gallon. The higher the price, the higher the taxes, it was the perfect storm and driving would never be truly affordable again, if it ever really was.

We can rewind to Chancellor Norman Lamont's Budget of 1993 to see where the complicated modern fuel tax system comes from. As well as increasing duty by 10 per cent it was the "fuel duty escalator", that raised duty by 3 per cent above inflation which has caught every bureaucrat's imagination ever since. This fuel escalator forced prices up from one of the lowest in Europe to one of the most expensive. When it was first added, fuel prices rose by 3 pence a litre and tax contributed to 72.8% of the total cost. In November that year, the new Chancellor Kenneth Clarke pumped the escalator to 5 per cent above inflation. When the government changed in July 1997, the new Labour Chancellor Gordon Brown boosted the escalator rate to 6 per cent above inflation.

The ever-rising cost of a litre of fuel meant that finally the British nation cracked. It was the professional drivers who led the charge so it was mainly the road hauliers and farm workers. As a result, go-slows through towns and on A roads as well as blockades of oil refineries were all good fun, but the fuel duties remained the same.

However, the Pre Budget Report of November that year promised reductions in duty for ultra-low sulphur petrol and conventional unleaded petrol of 2p per litre, plus a freeze on duty until 2002 and was implemented in the 2001 Budget. Then by October 2003, the Government was increasing fuel duties in line with inflation in the 2004 Budget.

In his first Budget as Chancellor in 2008, Alistair Darling announced that a 2 pence per litre increase expected to be introduced on 1 October 2008 would be postponed, leaving fuel duty rates at 50.35 pence per litre. Other planned increases were postponed as the Chancellor admitted that the global credit crunch had increased pump prices and wanted to help.

Then the motivation for rises started to change. A two pence per litre increase was introduced on 1 September 2009 with plans for a one pence per litre increase each year from 2010 to 2013, which the Chancellor claimed would save two million tonnes of CO2 per year. In Labour's final budget it was announced that rises in fuel duty for 2010 would be staged, with a one penny per litre increase on 1 April and 1 October and there was a further increase of 0.76 per litre on 1 January 2011.

Just to prove that governments are all pretty much the same, the incoming coalition continued with the planned increases. That was in their first Emergency Budget of June 2010. However in their first full budget fuel duty was cut by 1p per litre from 23 March 2011. Not only that, the inflation rise was postponed until January 2012, and the April 2012 inflation rise extended to August 2012.

Chancellor George Osborne decided that the fuel duty escalator that adds a penny above inflation would be cancelled. A new petrol tax buzz phrase 'Fair Fuel Stabiliser' was introduced for the March 2012 Budget, and above inflation rises would return only when the oil price fell below $70 a barrel.

The important thing to remember is that the rate on petrol and diesel had been frozen since December 2011 at 57.95 pence per litre. (In December 2008, fuel duty rose from 50.35p to 52.35p a litre. In April 2009, it rose again to 54.19p. In September it was increased yet again to 56.19p and then to 57.19p in April 2010. After the election, the coalition government increased it once more to 58.19p in October and then 58.95p in January 2011. Finally, in April 2011 it was cut to its current level of 57.95p.)

This level of tax was also applied to biodiesel and bioethanol. The rate on liquefied petroleum gas (LPG) was 31.61 pence per kilogram. The rate on natural gas such as biogas, was 24.70 pence per kilogram; and the rate on 'fuel oil' used for heating is 10.70 pence per litre. VAT was applied after fuel duty.

"Red" diesel, used in marine engines and agricultural vehicles, was more lightly taxed at 11.14p per litre. As we all know, it's illegal to use red diesel to power a road-going vehicle. Liquid petroleum gas was charged at 31.61p a kilogramme, although according to the AA, LPG usage was limited because grants were no longer available to convert vehicles and no LPG variants were manufactured in the UK.

It can be argued that such a high level of duty distorts economic patterns of consumption and production and even prevents workers from getting to jobs, which can only be reached by private transport. Economically it is regressive and results in lower income levels. Meanwhile, anyone living in rural areas, cannot be persuaded to take public transport by high duties in the way that the environmental lobby demands, because there just aren't any buses or trains in the countryside. A car is a necessity for rural dwellers, not a luxury.

By way of mitigation the Coalition government stated that as part of its commitment to help with fuel costs in remote rural areas it was considering introducing a fuel duty discount in such areas. Subsequently the Chief Secretary to the Treasury, Danny Alexander, announced in October 2010 that a pilot scheme would be introduced that would deliver a maximum of 5 pence per litre duty discount on petrol and diesel in the Inner and Outer Hebrides, the Northern Isles, and the Isles of Scilly. Indeed, this scheme would be extended until 31st October 2023.

Clearly governments can listen and make localised exceptions, so perhaps all rural drivers who have no access to any form of public transport should be given these concessions.

Fuel tax good or bad? Well it is bad that we have to pay it, but effectively it is always too high and therefore economically damaging and socially unhelpful.

High fuel tax affects consumption and production. Delivering goods or a service can become prohibitive or just too expensive. A business could fail because of that. When it becomes expensive for people to get jobs, in rural areas and especially shift work because of the cost of travel. Although much is made of charging drivers for using their cars, that is precisely what fuel duty does. Don't use the car and it is zero. If you live in a sparsely populated area with poor public transport access, then it should be lower. There are alternatives, except they aren't used as alternatives, rather they are additives, more tax. Indeed, we do reluctantly go into Ultra Low Emission zones and Road Pricing. When we did though, we were charged.

We can't leave it there, because they, the government,

are always looking for new excuses to raise car tax. Thank your lucky stars that you probably don't live in Jersey. I believe it is a bit more sunny there, and if you are not living there for low tax reasons then you probably can't afford the proposed 6p fuel tax rise.

The rise is part of the Jersey Government plan aimed at raising money to help combat climate change. The Environment Minister, Deputy John Young, said that the increase is designed to change islanders' behaviour. In 2019, the States declared a climate emergency, and pledged to become carbon neutral by 2030. More on this madness in a few chapters time.

Here's a fantastic chart by Peter Edwardson who kept a record of petrol prices from 1983 at speedlimit.org. He told me he should have done the same for beer prices.

According to Peter, "over the 36-year period, they haven't really moved much in real terms, but in the five years up to 2000 there was a 50% increase in the pump price, thus triggering the fuel protests. This was the only event that really rocked the Blair government, and could be said to have presaged the Brexit vote sixteen years later.

"The Conservative-led governments of the 2010s have effectively defused this issue by freezing duty for nine years, which is perhaps an achievement that hasn't received the recognition it deserves."

| Petrol Prices 1983-2019 | | | | | |
|---|---|---|---|---|---|
| Year | Price per Litre (p) | Price per Gallon (£) | Retail Prices Index | Petrol Price in constant terms (1983=100) | 5-year % increase ¶ |
| 1983 | 36.7 | 1.670 | 83.1 | 100.0 | - |
| 1984 | 38.7 | 1.759 | 87.5 | 100.0 | - |
| 1985 | 42.8 | 1.946 | 92.8 | 104.3 | - |
| 1986 | 38.2 | 1.737 | 96.7 | 89.4 | - |
| 1987 | 37.8 | 1.719 | 100.6 | 85.0 | - |
| 1988 | 34.7 | 1.578 | 104.1 | 75.4 | -5.5 |
| 1989 | 38.4 | 1.746 | 112.3 | 77.4 | -0.7 |
| 1990 | 40.2 | 1.828 | 121.4 | 74.9 | -6.1 |
| 1991 | 39.5 | 1.796 | 131.4 | 68.0 | 3.4 |
| 1992 | 40.3 | 1.832 | 136.7 | 66.7 | 6.6 |
| 1993 | 45.9 | 2.087 | 139.3 | 74.6 | 32.3 |
| 1994 | 48.9 | 2.223 | 133.1 | 77.6 | 27.3 |
| 1995 | 50.9 | 2.314 | 147.5 | 78.1 | 26.6 |
| 1996 | 52.9 | 2.405 | 151.5 | 79.0 | 33.9 |
| 1997 | 57.9 | 2.632 | 155.4 | 84.3 | 43.7 |
| 1998 | 60.9 | 2.769 | 160.8 | 85.7 | 32.7 |
| 1999 | 61.9 | 2.814 | 164.1 | 85.3 | 26.6 |
| 2000 | 76.9 | 3.496 | 168.4 | 103.3 | 51.1 |
| 2001 | 77.9 | 3.541 | 173.1 | 101.8 | 47.2 |
| 2002 | 69.9 | 3.178 | 174.5 | 90.6 | 20.7 |
| 2003 | 77.9 | 3.541 | 179.9 | 98.0 | 27.9 |
| 2004 | 77.9 | 3.541 | 184.6 | 95.5 | 25.8 |
| 2005 | 79.9 | 3.632 | 190.5 | 95.0 | 3.9 |
| 2006 | 88.9 | 4.041 | 195.0 | 103.2 | 14.1 |
| 2007 | 87.9 | 3.996 | 204.4 | 97.4 | 25.6 |
| 2008 | 103.9 | 4.723 | 212.1 | 110.9 | 33.4 |
| 2009 | 89.9 | 4.087 | 211.3 | 96.2 | 15.4 |
| 2010 | 111.9 | 5.087 | 220.7 | 114.8 | 40.1 |
| 2011 | 129.9 | 5.905 | 232.5 | 126.5 | 46.1 |
| 2012 | 134.1 | 6.096 | 240.8 | 126.1 | 52.6 |
| 2013 | 138.9 | 6.314 | 248.7 | 126.5 | 33.7 |
| 2014 | 130.9 | 5.951 | 254.8 | 116.3 | 45.6 |
| 2015 | 109.9 | 4.996 | 257.1 | 96.8 | -1.8 |
| 2016 | 103.9 | 4.723 | 261.1 | 90.0 | -20.0 |
| 2017 | 117.9 | 5.360 | 269.3 | 99.1 | -12.1 |
| 2018 | 115.9 | 5.269 | 278.3 | 94.3 | -16.6 |
| 2019 | 119.9 | 5.451 | 285.1 | 95.1 | -8.4 |

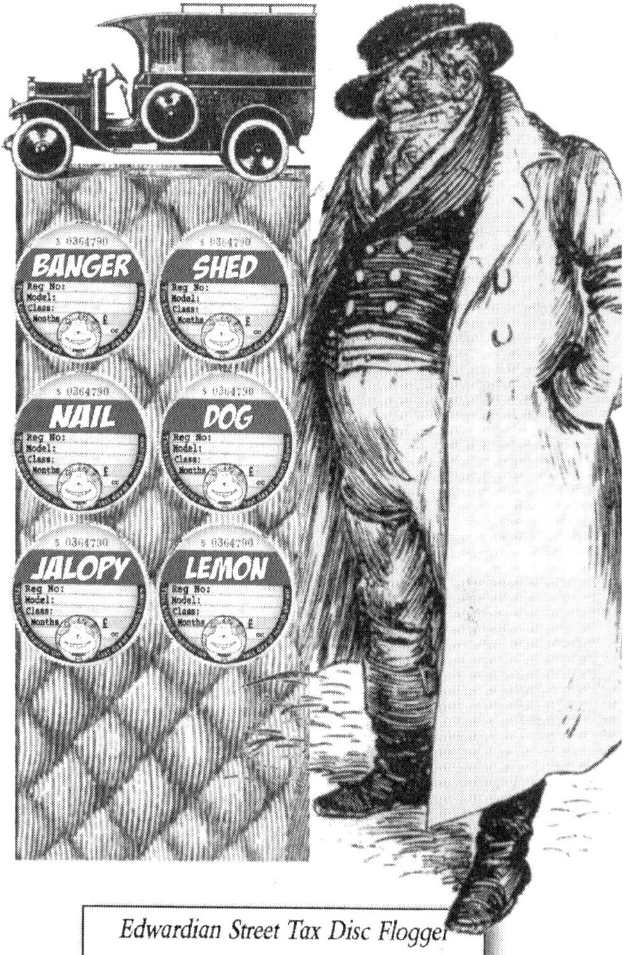


*Edwardian Street Tax Disc Flogger*

## 4. Road Fund Myth - *If it moves and even if it doesn't, tax it.*

There is no such thing as Road Tax, or even the Road Fund Licence. There is simply tax applied to vehicles, which is collected and then redistributed to be spent on just about everything. Not roads judging by the current depth of potholes in the UK. Of course it wasn't always like this. Unfortunately we will have to wade through lots of numbers, but that's all tax is anyway.

Originally road building was all paid for by ratepayers and the Tram companies due to the Tramways Act of 1870. Drivers have been taxed ever since the first powered vehicles appeared on the road. From 1889 owners of vehicles paid two guineas in vehicle excise tax to the Inland Revenue. The first petrol cars were legally defined as "light locomotives" in 1896, so were subject to a charge.

Then the Motor Car Act of 1903 introduced a £1 registration fee for any vehicle, Those used as taxis had carriage duty and that was £2 2s for cars of up to 1 ton. Then things became rather more complicated.

The Finance Act 1909-10 taxed owners on the Royal Automobile Club (RAC) devised formula according to their horsepower rating, which was referred to in the old days as the 'Treasury Rating'. It was imposed on a sliding scale ranging from 2 guineas to £42 each year. From 1921, the Government simplified this to £1 per RAC horsepower. This was reduced to 15s. per horsepower in 1935, then rising to 25s in 1940. If the war was making petrol difficult to find, private motorists were also being priced off the road, but all in a very good cause.

The authorities need to keep tabs not just on the driver, but also what they are steering. The Roads & Finance Act of 1920 required councils to register all vehicles, allocate a separate number and charge them direct. To make this easier, displaying a tax disc on the windscreen became compulsory in 1921. The Road Fund was established in the same year as the Ministry of Transport became responsible for collecting and spending the money. This Road Fund Licence was paid directly into the Fund and used solely for road construction until 1937. After that the 1936 Finance Act abolished the Road Fund element so that vehicle taxation went into a big pot, better known as the Treasury.

An interesting observation from the tax disc anorak community is that from 1923 to 1938 the road tax disc security background text read Road Fund Licence. Then in 1939 the tiny text switched to, Mechanically Propelled Vehicle Licence. This signaled that the link between paying a specific amount to maintain the roads was gone forever.

Car Tax was fiddled with in 1947 because keeping a cap on the horsepower of British cars made them underpowered and less attractive to the export market. Taxation was now linked to engine capacity, at the rate of £1 per 100 c.c. There were still anomalies and pre 1946 cars were treated differently so in 1953 all cars were subject to a flat-rate tax of £12 10s. There were also quirky exceptions if you could be bothered to look for them, which meant that very old and tiny cars with engines that did not exceed 7hp paid even less.

Car tax went on being car tax, the Vehicle Excise Duty (VED) euphemism we will ignore. It is a tax on owning a car, so we will refer to it as Car Tax from now on. Now, let's try and get excited about the administration and the bureaucrats involved.

Vehicle licensing went from being a local government administrative matter to a national one, so the decision was made in 1965 to centralise it, by taking advantage of what Prime Minister Harold Wilson called the 'White Heat' of new technology, namely data processing. In consultation with the Department of Employment, so clearly this was also a job creation scheme, the Driver and Vehicle Licensing Centre (DVLC) was opened in Swansea. There were still regional offices though, 81 in all around the country and there was also plenty of vehicle related form filling you could do through the Post Office. In April 1990, the DVLC became an Executive Agency of the Department of Transport known as the Driver and Vehicle Licensing Agency, DVLA. Don't you love initialism?

The tax disc was discontinued in October 2014. There was newer and apparently cleverer technology to replace that bit of paper. Number-plate recognition cameras allowed the authorities to check electronically whether car tax had been paid. Relying on a network of Automatic Number Plate Recognition (ANPR) cameras on the roads or in police cars. This system exchanges data with the DVLA database, which keeps a record of all taxed and untaxed vehicles. The theory was that getting rid of the paper tax would lead to a tax windfall and make the government look good as they sort of became a monolithic paperless office, like those consultants promised, also making the government efficient and digital.

The reality is that tearing up that groovy little disc actually costed the government and therefore the taxpayer, well over £100m. Tax evasion seems to be a lot easier now the disc has been dropped. Untaxed vehicles are quite often not roadworthy, so won't have an MOT and the drivers will probably regard insurance as a formality as well. According to the Department for Transport (DfT), 1.8 per cent of vehicles on UK roads in 2017 were unlicensed which was a rising figure and amounted to almost 800,000 cars and commercials. Losses due to tax evasion were £35m, and then by 2018 it had shot up to over £100m.

Something as basic as a visual reminder clearly works. A disorganised motorist genuinely forgets and a dishonest one sees an opportunity to dodge the tax. At least in the old days someone slipping a Guinness beer label into the holder showed some creativity.

Since the changes it has not been possible to transfer

any unexpired tax to the new registered keeper, meaning a new owner must tax the car before driving away. Similarly, when selling the vehicle, the DVLA has to know. So the government makes a guaranteed £30 profit every time a car changes hands because sellers are only refunded whole months and the buyer is charged for tax from the start of the month. Some may think that is a reasonable admin cost, but that is still tax payer money and this is the most basic of double entry book keeping procedures.

For a long while car tax was just that, a flat tax applied to all models irrespective of their size or engine capacity. Clearly that situation could not be allowed to continue. Indeed, going right back to the good old fashioned Treasury Rating was going to be a great way to boost car tax revenue. At this point we could compile loads of boring charts, but it better to highlight how radically the Government restructured tax which on the face of it, became fairer. Now the CO2 polluter would pay.

We deal with CO2 elsewhere in this book. Certainly charging smaller engined cars less would be far better and if the lower CO2 (as found on the V5C document) was one way of getting a lower tax rate, then that will have to do.

Vehicle tax rates are based either on engine size not over 1549cc and over 1549cc, or fuel type and carbon dioxide emissions, depending on when the vehicle was first registered. However, for cars registered on or after

1st March 2001, the rates were split into bands – the lower the emissions, the lower the vehicle tax. A car with CO2 emissions below 100g/km was road tax-free. That would not last, as now only cars with 0g/km CO2 emissions are exempt from paying tax.

The flaw with this system is that the older pre 2001 cars go on polluting whilst the later ones will pay near £500 a year just because their emissions are slightly over 225g/km. That means used examples could eventually end up being worthless because they can cost more to tax than their notional trade value. Seems odd that, but it remains a loophole that allows car enthusiasts to buy interesting large engined cars from before 2001 and not be penalised for it.

However, as revenues have fallen because of the CO2 bands and more manufacturers offering low emission cars, in came the 'showroom tax'. Buyers of new cars registered from April 2010 incurred a first year tax rate. Except that there was no first-year tax to pay on cars in sub 130g/km band vehicles. The standard rate subsequently switched to a flat fee for all cars except those emitting zero emissions, such as electric vehicles and some hybrid vehicles. There was also an 'Envy Tax' with an additional £310 per year for all cars with a list price of over £40,000.

Meanwhile, tax on alternative-fuel cars including hybrids, plug-in hybrids, and those that run on LPG, CNG or biofuel was fixed at £10 less than regular petrol or diesel cars.

Can you avoid paying car tax? Of course you can, but the more important question is, should you? Let's say you decided to make some sort of daft protest and not pay the car tax then you'll get a Late Licensing Penalty. You'll get a fine and if you are caught with the untaxed car you'll get an Out of Court Settlement letter. That's a fine plus one-and-a-half times the outstanding vehicle tax rate. Then if you don't pay it, this becomes a magistrate's court matter. Plus the fine goes up. There is a chance that your car could be clamped and if you don't pay the fine and get it released then it could be impounded and ultimately auctioned or crushed. Or both.

It is an escalating cycle of non-payment hell and the only way to avoid it in the first place is to make a Statutory Off Road Notification (SORN). The thing is though, it has to be off the road, parked somewhere on private property, say as an art installation in a sitting room. Ideally a garage and not squished onto the frontage of your terraced property, which could trigger neighbours into doing something drastic.

Then there is the living in the past option courtesy of the Historic vehicle exemption. Cars that are over 40 years on January 1st won't pay any tax. It is just a case of finding a car, or cars that you can live with on a daily basis that are so old.

Alternatively the Government could put car tax on fuel. That old one has been around for decades. Pay as you drive right there. Drive zero miles pay nothing, drive

100K miles, and pay gazillions. At the moment this separate car tax system gives bureaucrats the opportunity to tax vehicles twice, make it all very complicated and keep them in work. There are also thousands of non-government employees like accountants who also benefit from a needlessly complicated tax system. Putting more tax on already overtaxed fuel is the simple answer, it also shows how overtaxed the British motorist actually is. There is also the small matter of working out how to tax the future deluge of electric vehicles. Maybe Government will just have to make all electricity delivery systems more expensive.

Back in 1974 Dennis Healey considered abolishing Car Tax. He didn't because of the potential loss of tax flexibility and the potential impact on the UK motor industry. One of the arguments against abolition was that it would favour smaller vehicles, which would generally be imported. But that was back in the old days when this country had an indigenous car industry, before it became globalised and exclusively foreign owned. The politics of building and selling cars continues with little consideration for the actual motorists.

Government likes Car Tax because it reinforces the whole edifice of vehicle licensing, and paying for its enforcement. It also plays a vital role in ensuring compliance with MOT and insurance certification and even discourages the ownership of second cars. That's right it makes everything complicated and certainly with heavy goods vehicles means more money to pay for any tarmac damage caused. Except that there seems to be little evidence that enough is ever spent on our roads. Also, Car Tax never addresses Heavy Goods Vehicles HGVs coming into the UK from Europe. The Government did introduce a £10 a day charge, which the EU disputed in court as unfair. Fuel tax wouldn't help either as they fill up before arriving here. Also HGVs can be registered and fuelled in Europe by British companies.

For environmentalists Car Tax can never be high enough, as they believe motorists never pay the full environmental costs of their activities. "UK transport policy is standing at a crossroads. The decisions the current Government takes on transport to tackle the dual challenges of climate change and rising oil prices could have significant repercussions for many years to come...Friends of the Earth is calling on the Government to: Change direction on transport policy - and aim to rapidly move towards a low-carbon transport system...Vehicle Excise Duty must be changed to make road tax on gas-guzzlers more expensive - and cheaper for greener cars..." Friends of the Earth; from briefing on 'Fuel prices, transport and climate change' – 2011.

You may think that there is nothing to look forward to, but road charging could well be the future and we explore the pros and cons of that elsewhere.

Whilst we are in the unfair car tax area, there is the small matter of, Insurance Purchase Tax. This fairly terrible tax was introduced to the UK by Kenneth Clarke in the 1993 Budget. It was set at 2.3%, but of course it didn't stay at that level. By 2011 it was 6%; by 2015, 9.5%, 2016, 10% and in 2018 12%. At the last count it was pulling in just under £5bn annually, a massive income for the Treasury. It isn't just applied to cars of course, but with over 32.5 million cars on the road, that's a huge chunk of the revenue. This is a good example of a regressive tax, just like car and fuel duty, as it takes a larger percentage of income from those who have less money to spend. For example young drivers are worst affected, because they are a much higher risk. So a big premium with even more insurance tax on top.

At the time of writing Demotorized, there was still something called the BBC Licence fee (although it was going to be reviewed), which is compulsory. Don't pay it and you end up in prison. All that licence money is as some people love to say, 'ring-fenced ' and given to TV presenters, idiot disc jockeys and the duplication of programmes and services that are produced commercially and arguably better elsewhere. Simply spending car tax on the actual road infrastructure that creates wealth for the country would at least be an investment.

Telly Tax good. Car Tax bad. Couldn't agree less.

Things were never great for the motorist under any

government. The arrival of New Labour in May 1997 did not mean everyone was humming 'Things can only get better' along with Tony, Cherie, Gordon, John and Mandy. That's because it didn't take long for New Labour in general and Deputy Prime Minister John Prescott in particular to suggest that not only was one car bad, but two would virtually become a criminal offence. They had declared war on second car families by promising to make public transport so good that we would abandon our cars. However, details about the double decker/Inter City carrot were sketchy whereas the stick being used to beat the poor old motorist (road pricing, car tax rises) was the size of a motorway flyover reinforcement.

I was reminded by a Mr. J. Clark of Oxford that the incoming coalition government of 2010 promised quite a lot. Philip Hammond, the new Transport Secretary said, "We will end the war on motorists. Motoring has got to get greener but the car is not going to go away." Despite keeping to some pledges like outlawing rogue clamping companies, they remained anti-motorist business as usual.

*SPECS vs GATSO*

## 5. Speed Pays - *Milking Motorists for Money since 1896.*

In the first half of 1896 back when driving still wasn't strictly legal, Walter Arnold from East Peckham, Kent became the first person in the world to be charged with a speeding offence. For a start he was on board a locomotive and there was no horse involved. By law he should have had three extra mates on board and rather than a registration plate, which hadn't been invented yet, his name and address should have been attached to the vehicle. He could have got away with 2mph, but went all out at 8mph, hence the court case. The Magistrates bench applied a 5-shilling fine plus costs.

Getting rid of that bloke with the red flag was the first stage in helping those early powered vehicles to make progress that was above walking pace. As has already been explained, the red flag requirement had been done away with many years before, the whole flag-bloke-walk-thing was just a symbol of how idiotic the restrictions were.

Not surprising that this monumental change was celebrated in the Emancipation Run from London to Brighton in November 1896 which as we have already mentioned involved some flambe' de flag. Later it became known as the London to Brighton Veteran Car Run, open to cars built before 1905, the event was nostalgically re-enacted in 1927 and has been held nearly every year since. The significance of the run may be largely forgotten, as these days it just looks like a load of ancient, noisy cars, sometimes with celebrities on top of them, in fancy dress, trying not to break down on the A23, or expire of hypothermia.

Back when Walter Arnold was tearing around Kent, motor vehicles, or locomotives, as the magistrates preferred, were limited to 4mph maximum, down to 2mph in residential areas. That's a speed limit, which would please campaigners today. By 1903 the maximum had been raised to a heady 20mph. By the 1930 the Road Traffic Act, speed limits for cars were abolished altogether. It was a golden age for freedom and accidents, but at least the roads were relatively uncluttered. It wasn't a complete free for all as there were clear penalties for reckless, careless or dangerous driving, plus driving under the influence of alcohol or drugs.

Mind you, most cars struggled to reach any sort of maximum speed. The roads were not great and in 1927, the Treasury believed that the main supporters of the Road Fund were private motoring organisations who wanted road improvements not for the good of the country but to drive faster: "it is clearly absurd that the State should be asked to provide large and ever-increasing sums for what are virtually pleasure racing tracks."

So although you could drive like a loon everywhere, by 1934 a new limit of 30mph was set for built-up areas to calm things down a bit. Provided drivers did things safely, they were trusted to go about their business without too many restrictions. A golden age? Well, for the few who could afford to buy and run a car, perhaps it was. Treating drivers as civilized grown ups seems like such a novelty now, largely in the sort of cohesive British society we had back then, it worked.

There is one politician who bestrides the 1950s and '60s like a colossus and burns even more brightly than Leslie Hore-Belisha, the Minister of Transport who in 1934 added beacons to pedestrian crossings, which were thereafter called Belisha Beacons. Also, those crossings were later painted in black and white stripes and became known as zebra crossings. Yes, more influential than him. A man who initiated changes around motoring and transportation that are still being felt today and resonate throughout so many of the Demotorized chapters.

There really should be books, documentaries and major motion picture films about the life and good times of Mr. Ernest Marples MP. Strangely there are few credits, no biographies, just the usual Parliamentary reports in Hansard. Overall he was a bloke from humble beginnings who made it really big. Some have called him a scoundrel and a bit of a bounder. Indeed, in 1975 and by that time a Baron, he left the country under a cloud of tax evasion. He died three years later, appropriately enough, in opulent surroundings in the tax haven that still is, Monaco.

Marples opened the first major section of the M1,

Britain's first city-to-city motorway, from near Watford to near Rugby in November 1959. He made a speech, "On this magnificent road the speed which can easily be reached is so great that senses may be numbed and judgment warped. The margin of error gets smaller as speed gets faster. New motoring techniques must be learnt. So here are my two suggested mottos: First: take it easy, motorist! And second: if in doubt, don't!"

Wise words there from the new Minister of Transport who also happened to be the majority owner of Marples, Ridgway and Partners, a Westminster-based civil engineering contractor. Odd then that he was responsible for building roads as well as cutting the ribbons on brand new ones. Even odder that he commissioned Doctor Beeching's report published in 1963, which was responsible for huge cuts to the rail network. Private car use was massively increasing because it was convenient, becoming more affordable and those new fangled motorways certainly helped. Beeching's report noted that half of the 7000 railway stations generated only two per cent of rail traffic and that one-third of the track carried only one per cent of the traffic.

Less well known is that Marples also commissioned the Buchanan Report of 1963, when civil engineer and planner Professor Colin Buchanan (Traffic in Towns) recommended that nothing be done at all to encourage urban cycling. Better to keep on building roads for cars. There again, cycling which had been a huge sport and practical way of getting around was falling out of favour. People actually wanted the comfort and convenience of a car rather than going everywhere on a pushbike. It really was the age of the car.

Marples was called out on his road building conflict of interest in 1960 and was forced to sell his shares. Some believe they went to his Mrs. and quite possibly into an offshore tax-free trust. The thing is that Marples wasn't pilloried at the time for Dr. Richard Beeching's recommendations, but motorists didn't like the yellow lines, traffic wardens and drink drive clampdowns which arrived on his watch.

Getting back to the first motorways, starting with the Preston Bypass (which later became part of the M6) in December 1958, they were still unrestricted. Here we have the real 'Harvest Gold Metallic' age as car ownership increased and vehicles became more capable. Then one fateful day the British motorist got an unwelcome early Christmas present when on 22nd December 1965 a temporary maximum speed limit of 70mph was introduced on Britain's motorways. The experiment initially lasted four months but was extended and then made permanent in 1967.

Legend has it that racing driver Jack Sears, and a 4.8-litre Ford-powered Cobra Coupé GT was clocked at 185 mph, the highest speed ever recorded on a British motorway. Sears needed a stretch of road long enough to max out the car in preparation for the Le Mans 24-hour race. This was not illegal, or unusual, as many car companies used the fairly traffic free motorways as unofficial test tracks. That wasn't the problem.

It was number of car crashes during the foggy autumn of 1965, which led to the Conservative government to hold consultations with the police and the National Road Safety Advisory Council. They concluded the crashes were caused by vehicles travelling too fast for the conditions. This still happens today, probably more often, so the blame is on idiot drivers. There was a proposal that a speed limit be used during periods when the road was affected by fog, ice or snow. Not only that, an overall maximum speed limit of 70 mph should be tested out. That trial period went on for a while and was turned into law by Labour's Transport Minister Barbara Castle in December 1965.

After that, speed limits were only seriously modified in order to save fuel when the first major oil crisis hit. A voluntary limit of 50 mph on all roads was made statutory in December 1973 and stayed in place until the end of March 1974, when it was lifted to 70 mph for motorways. The limit was subsequently increased for other roads in May but this proved only temporary as limits were reduced again in December as part of a package of energy saving measures. The new limits were 70 mph on motorways, 60 mph on dual carriageways and 50 mph on single carriageway roads. The limits were revised again in 1977 to what we have at the time of Demotorized was published. That could of course always change.

Despite improvements in vehicle safety and performance, 70 mph was as fast as British drivers could legally go. After that it was cat and mouse with coppers in Jags and Rovers chasing down speeders on motorways, but essentially driver's always had a sporting chance. Significantly there were a lot more patrol cars and policemen and Traffic Policemen around. Even so, they could not be everywhere at all times of the day and night to catch motorists who were not just speeding, but jumping red lights and even straying into bus lanes. The Road Traffic Act 1991 changed all that. From then on evidence from type-approved automatic devices could be used as the sole evidence that a motoring offence had been committed.

The first speed camera in 1991 was located on the M40 motorway in West London. The cameras used rolls of film, which had to be developed, how quaint. Legend has it that this camera used up its first 400-exposure roll within 40 minutes. Meanwhile, The first trial Gatso Camera (see further) was fitted on the westbound section of the A316 over Twickenham Bridge in Surrey. A dual carriageway with a 40mph limit, but incredibly the camera was set to snap at 60mph. They caught a thousand speeders a day in the first 22 days according to the Richmond and Twickenham Times. A Home Office study in 1996 claiming that areas covered by the cameras had seen a 28 per cent reduction in injury crashes at speed camera sites and an 18 per cent reduction in injury crashes at traffic lights meant that these devices were here to stay.

Clearly there was some education to be done. Taking incriminating pictures of a speeding driver would undoubtedly save lives at a notorious traffic black spot. Unlike other methods of road traffic law enforcement, speed cameras do not require offending motorists to be pulled over. They simply receive a fixed penalty notice. Effectively this is hands off policing conducted by the Royal Mail. However, it did seem to be working.

Home Office figures showed that by 2000 around 600,000 motorists were caught speeding by cameras in England and Wales. In Northamptonshire for instance, accidents fell at fixed camera sites by 50 per cent and the number of people killed and seriously injured by 67 per cent in 2001. This is a significant year. That's when local authorities were allowed to get some of the camera income. Police and councils teamed up to form safety camera partnerships The National Safety Camera Scheme was included in the Vehicles (Crimes) Act 2001.

Not surprisingly, the number of cameras shot up to 2000 and more than doubled again according to figures from the AA. With lots more cameras around, the motorist became far more compliant and the number of camera flashes dropped. As a result the speed thresholds were deliberately reduced to restart the income stream. It was becoming more difficult to convince motorists that this was a safety, rather than a revenue issue.

To counter the public criticism and suspicion fresh guidelines were issued by the Safety Project Camera

Board in 2002. New rules required cameras to be painted yellow, not hidden behind trees for instance and be clearly signed. Also cameras could only to be sited at locations where four people or more had been killed or seriously injured in accidents, or eight personal injury accidents had taken place in the last three years.

In 2004 a Department of Transport report found that none of the 6,000 cameras across England and Wales were sited for revenue raising reasons. Well they would wouldn't they?

Nevertheless, the Transport Secretary announced the ending of the National Safety Camera Programme and netting-off funding arrangement for cameras in England and Wales. Nevertheless, local authorities, the police and other local partners were still given responsibility for the future deployment and operation of cameras, so nothing really changed. The money raised was just redistributed within a new administrative organisation. At least in 2011, speed camera "transparency rules" meant local authorities must publish the number of accidents at camera sites, before and after cameras were fitted.

Rather than concentrating on direct prosecution as the number of fixed penalty notices fell from 1.8 million to 1.1 million from 2007 to 2009, speed awareness courses became the new local authority revenue raiser.

Here is your get out of fixed penalty hell and penalty points, free card. All you have to do is pay for a half-day speed awareness course. However, you do need to be offered it first. If your speeding is excessive, or you are a repeat offender then probably not. Also the police force involved may not subscribe to the speed awareness shenanigans. If you do get the option then you'll pay effectively what the speeding penalty would be.

I've not been on a course, but have spoken to a lot of guilty attendees. These courses are run by independent consultants, well-known motoring organisations and even police constabularies themselves. Apparently it pays to be positive and not question any narrative. So they will go on about stopping distances and being aware of your surroundings, which is all useful stuff. They can show pictures and video of accidents and the consequences of not slowing down when you should. Indeed, a friend of mine was actually in one of the photographs of a tragic scene. He didn't cause it, but as a fireman was there releasing people from the wreckage. That provided the only light moment of the day. Essentially the course is like a bribe to behave better and that may well work for some.

These speed awareness courses are run under the umbrella of the National Driver Offender Retraining Scheme (NDORS) to ensure consistency across the 43 forces. One aspect where there is no consistency is the fee charged for the course. Even if all the money is ploughed back into road safety, drivers are still paying for what is a job creation scheme.

Back to the cameras, the funny thing is that many police

forces have admitted that there are rather a lot of empty yellow metal boxes out there. According to the AA in late 2017 only half of fixed speed cameras are actually switched on. Out of the 2,838 cameras around 1486 were active. Indeed Northamptonshire, Cleveland, Durham and North Yorkshire police all said that all of its cameras were shut down in 2011, but they remained in place as a deterrent.

That deterrent may work, but speed cameras have a margin of error or at least tolerance to bother drivers. Cameras may have a 10 per cent leeway plus 2mph. Mathematically that means that in order to get a ticket in a 30mph residential area you would have to hit 35mph. The National Police Chiefs' Council (NPCC) suggested that 10 per cent was guidance rather than being set in stone. However, each force had their own parameters and might reveal it, if you asked them nicely through a freedom of information request. Some didn't. This is the perfect example of the rigidly defined area of uncertainty that drivers continue to operate under.

The ultimate abrogation of responsibility is the empowerment of the local community to dress up in high visibility outerwear and then point guns at their neighbours. Say hello to the Community Speed Watch Scheme.

This is where volunteers use hand-held devices that tell them the speed at which drivers are travelling through their village, or residential area. When a driver is going a set amount above the speed limit, usually 10%, the vehicle's registration, make, model and colour are recorded. The speeder's details are then passed to the local police force, which could add them to a national database. Repeat offenders certainly will be followed up and at the very least get a letter from the local constabulary. However, no speeding tickets are issued, this is all about education and quite possibly settling local scores. Incidentally, they can only operate in daylight hours at locations approved by the police.

One rather worrying development occurred in 2019. Transport for London are very open about their Travel Safe Priority Areas trialed "to combat community concerns in certain inner London areas with high visibility and covert policing".

That is because there is a programme to lower the central London speed limit to 20mph as part of wider plans to eradicate death and serious injuries on the transport network by 2041. That is all very laudable, but it has to be questionable to use children to shame speeding drivers. These are called Junior Roadwatch sessions, where children get to wear a high visibility jacket and work with the police to educate drivers about the dangers of speeding.

Clearly, Big Brother (and little school age brothers and sisters) is certainly watching us, or at the very least that nosey so and so from next door. There may not be much that we can do about this, but maybe it is worth knowing what technology the authorities are using against the motorist.

**Gatso** The original speed camera dating from 1991, made by Gatsometer, hence the name. It's a rear-facing camera, so it can catch motorcycles as well. Painted yellow they can be found at the side of the road or on motorways on the overhead gantries. There are also road markings. You will only be caught driving away from it, never towards. Originally film, digital from 2007.

**Truvelo** Again, named after company that makes it. Like a Gatso on a pole or gantry, but it is forward facing. Takes a picture of the driver for identification purposes. There are confirmatory road markings too. Truvelo D-Cam is a digital version and can be mounted forward or rear facing and can also be used at traffic lights.

**HADCES 3** Highways Agency Digital Enforcement Camera System 3, designed for smart motorways. Smaller and harder to spot, but still on poles. Rear facing with confirmatory marks on the roads. It's party trick is that it checks average speeds between the poles. They can also catch drivers in the wrong lane.

**SPECS** Specifically an average speed camera. Often used in roadworks. Its main purpose is to utilise Automatic Number Plate Reading (ANPR) technology to register vehicles as they pass. The first camera logs the vehicle with a time and date stamp. Once the vehicle has passed the second camera, the time stamps on the two images are compared, and if the time taken to cover the distance means the average speed is higher than the limit, instant ticket.

**What's inside a camera van?** Mostly Police. It is marked on the outside as a safety camera vehicle, but it still looks like a Police van and they can film through windows or doors. Often parked in dual carriageway or A road laybys and bridges. Their mini Gatso cameras use radar technology but there are also handheld radar or laser gun cameras. They can be set up to catch speeders in either direction.

**How to dodge a speeding ticket** Legislation around speed limits refers only to motor or mechanically propelled vehicles. In that case get on a horse, a bicycle, or a human propelled scooter and knock yourself out.

Anybody ever fought back against all this? Well there was the self-styled Captain Gatso from Motorists Against Detection (MAD). They took direct action in a very Gallic way by targeting the cameras themselves, burning them to the ground, or just booting them over. You either regard this as mindless vandalism, or a blow for freedom from surveillance.

A few words about variable speed limits These can be set at busy times, in conjunction with a Red X sign on the display board. It probably makes more sense when used to manage an incident, or there is something in the road, effectively an unexpected item in lane 2. These speed limits can also be automatically triggered by sensors that monitor traffic flow. The really important thing to remember is that any speed limit displayed inside a red circle is legally enforceable. Everything returns to normal when the national speed limit sign is displayed again. This is supposed to be an upside of the smart motorway system. It is designed to improve traffic flow and stop bunching. Not at all installed to catch drivers out.

Clearly, Smart motorways just aren't. In the light of tragic accidents and fatalities, their safety for ordinary road users has always been in question. Giving evidence to the House of Commons Select Committee in 2019 the head of Highways England told them that so-called "dynamic" smart motorways, where the hard shoulder is used as a live lane during peak times, are "too complicated for people to use" and that the system would not be extended to other roads.

These Smart motorways were designed so that capacity could be increased and congestion reduced, rather than going to the expense of widening roads.

The "dynamic" design is in use on sections of the M1, M4, M5, M6, M42 and M62 accounting for 68 miles in all. There is 135 miles of the "all lane running" system, with its emergency refuge areas. There are also "controlled motorways" - which have variable speed limits and can be found on around 120 miles of motorway.

Is it any wonder ordinary motorists find them complicated and dangerous? Road users are uncertain how to use them, because their operation simply isn't logical, or safe. Indeed, I have been told off any record whatsoever that the brains behind these Frankenstein motorways went to work on the tube.

Speed Limit Statistics are not what they seem. The terrible thing is that the accident and injury figures are all lumped together. So the fatalities and injuries, however minor are all lumped together for effect. Should that matter? Well that isn't telling the truth, clearly it has a dramatic effect, but making something sensational rather than educational is bad.

The Alliance of British Drivers (ABD) has long campaigned for an independent body to interpret the accident statistics especially as they relate to speed. In 2016 1,800 people died on the roads and around 24,000, seriously injured. If 15% of the fatal and 7% of serious injury crashes are above the speed limit, how many of 217 deaths involve the sober, legal drivers travelling a few mph above the speed limit? That is who all the campaigns are aimed at rather than the inebriated, drug takers and irresponsible who may well be causing the majority of accidents. Sadly we will never know, until serious changes about the way the statistics are compiled is changed.

Rather excitingly it has been suggested that the maximum speed limit could be raised to 80mph. Just when drivers thought that they would be stuck at 70mph until cars are actually banned, there is possibly a chance of going 10mph faster. Fantastic, provided that you have an electric car. There are separate Demotorized Chapters on aspects of all this, but this issue unites it. The motorway top speed came under scrutiny back in 2011, when an independent climate change committee warned that an 80mph speed limit would generate an extra 2.2 million tonnes of CO2 a year. The thing is though if those speeding cars are mostly electric, then there would be zero pollution. Well, there isn't, but more on that later. Also these will be smart cars, ones that can have the technology to keep a safe distance, keep in the lane and probably be autonomous to the required level. That is why they are looking at the issue again

The problem with Speed Cameras is that for the majority of their existence they have been used primarily not as a means of increasing road safety, but of raising revenue by police forces. They can turn otherwise law-abiding drivers against the police and the Government because all of a sudden instead of concentrating on catching burglars, they go for the easy collar, on a Polaroid. Not only that, cameras penalise someone who goes over the set limit rather than a driver who happens to be dangerous, drunk is uninsured (APNR flags up, but won't pull over), or driving an unroadworthy vehicle. The film takes no account of the context of the speeding offence. Worst of all cameras are now part of the road furniture, which are unlikely ever to be dismantled, or uninvented. Speeding doesn't kill people, but bad driving does. What we need is higher driving standards. It is that simple.

With certain obvious exceptions perhaps every single

driver should qualify on two wheels first. The road awareness required to ride a moped or motorcycle safely is far in excess of that needed to steer a car.

Speed is a serious issue. We should all learn how to deal with it better.

*The 'All-New' Top Gear Team in 1886*

*No Parking in shop fronts, or on cats...*

## 6. Space Race - *Parking Means Business.*

Cars have to go somewhere when they are not being used. That could mean outside your house, place of work or in a garage to highlight just a few possible locations. Trouble is, where there is available space for some, there is a rental opportunity for companies and taxation exploitation for local authorities and government.

It was the Edwardians who first realised that neatly organising vehicles into a clearly defined, architecturally interesting buildings could be gainfully monetised. The City & Suburban Electric Carriage Company located in London at 6 Denman Street, opened the world's first multi-storey car park in 1901. The garage had seven floors, 19,000 square feet, space for 100 vehicles and an electric lift to move the vehicles around. The company opened a second building a year later that could hold 230 vehicles. One of the most interesting multi-car garages of that era was The Electromobile Garage, located in Carrington Street, London from 1907. As the name suggested it was mainly a berth for electric vehicles where they could be charged, cleaned and generally looked after. Indeed, 300 EVs stacked together over two floors which was years later owned by National Car Parks (NCP).

That's all very interesting because everyone is having kittens of excitement over what now is, the second coming of the electric vehicles. Significantly, the owner of that location owes its existence to the Second World War. The simple truth is that the biggest changes to parking in the UK came courtesy of Hermann Wilhelm Göring and his Luftwaffe over which he was Supreme Commander.

The facts are that NCP was founded in 1931 by Colonel Frederick Lucas, but things didn't really get going until he linked up In October 1948 with Ronald Hobson, who together with his business partner Sir Donald Gosling, founded Central Car Parks when they invested £200 in a bombsite located in Holborn, central London. They realised it was possible to take advantage of all the pre-flattened areas which could be converted cost effectively into car parks. With their expertise, NCP also managed sites on behalf of third parties.

If it bothers you, NCP isn't very British anymore. There has been rather a lot of City Hedge Fund action, which has seen it change hands many times since the late 1990s. Never mind the considerable income, the company had also become a massive property empire. Since August 2017, NCP have been owned by Park24, a listed Japanese company, and Development Bank of Japan. Park24 is the majority owner based in Tokyo with 1,043,570 spaces in over 19,000 sites, across 8 countries. This should bother you because when you park in an NCP on a little bit of painted rectangle that will be forever Britain, you are sending money Far Eastwards. Ultimately, that does matter to the balance of payments.

Interestingly, back in 1999 NCP created the UK's first private-public partnership, setting up a joint venture with Manchester City Council to establish NCP Manchester Ltd to manage car parks in the area. That agreement was renewed in 2019, which is all rather frightening, if a ruthlessly efficient way of operating parking spaces. The power however, has in recent years been with the local authority as well as the police and the Traffic Wardens they controlled. It was the Road Traffic Act 1984 which enabled councils to enforce particular parking regulations and the Road Traffic Act 1991 extended their responsibility to yellow line matters.

Taking advantage of space and turning it to commercial advantage In the UK this is where the 'parking industry' came from. Initially it really was a case of plonking a tiny shed on a bombsite to accommodate an old soldier with a war wound limp to take the money from motorists for a daily or hourly period. After that, commercial car parks seemed like an excuse for some inappropriate 'Brutalist' concrete architecture in City centres.

This 'industry' became an even bigger part of the establishment when the British Parking Association was formed in 1967. Its members came from all the important parking related areas including Local Authorities, with the land, Development Boards with their eyes on the bottom line, Parking Operators who were already in the game, Equipment Manufacturers who were coming up with clever ways of marking time and space, plus the people who would ultimately do the heavy lifting such as Contractors, Engineers and Architects.

The single most important personality when it came to parking was our old friend Ernie Marples. Under his very long watch as Minister for Transport (1959-64) an awful lot happened to make parking more of a challenge for the average British motorist.

Road markings have been around for quite some time, in particular the humble white line since 1918. Instructional white lines have at least kept everyone on the right side of the road, which in the UK means the left. These were mainly used to stop drivers at junctions and were only standardized in 1926. Yellow lines were not introduced until the 1950s, and designed to communicate restrictions on parking, waiting, and loading. They were first seen in 1956 to deal with illegal parkers in the China Town area of Soho, London.

You can't blame motorists for wanting a decent meal back in the '50s, but clogging up the streets didn't go down well with the Metropolitan Police. It wasn't just about ordering a takeaway, although back then it was predominantly sit down meals. These yellow related rules and regulations also covered waiting, loading and unloading. Just because you were in a van or lorry, there was no excuse for outstaying your welcome. Then the Road Traffic Act 1960 changed everything

Not only were yellow lines doubled, there were now

wardens of the traffic. Yes, the Ladies and Gentlemen of the Traffic Warden army were officially only there to help drivers find a parking space and were certainly not policemen. They could though hand out two quid fines. History records that the recipient of the first ever ticket was Dr. Thomas Creigthon who carelessly parked his Ford Popular when answering an emergency call for a heart attack. The great British public was having none of that and the fine was later rescinded.

It would be an understatement to say motorists never warmed to Wardens, although The Beatles, 'Lovely Rita Meter Maid' humanised them brilliantly.

Paying for space became a council matter on June 10th 1958 when the UK's first parking meters were planted in Mayfair. Roughly 700 yards of the things were installed by Westminster City Council. An hour of being stationary cost 6 old pennies. Not that it took pennies, you had to shove a whole silver sixpence in the slot. As the first parking meter was unveiled in Grosvenor Square, George Nugent, the then parliamentary secretary to the Ministry of Transport, chose to quote Milton, to honour their installation: "They also serve who only stand and wait."

Fighting back against the rise of these machines amounted to jamming them with coins wrapped in paper or folded matchboxes. Alternatively, it was possible to knock their tops off and keep the change

Hardly surprising that by 1963 motorists' were

campaigning against Ernest Marples and his oppressive Ministry of Transport dictatorship. Their tactics were fairly extreme, downright rude and possibly blasphemous. Yes, 'Marples Must Go' car stickers are as bad as it got and most famously that slogan appeared on a motorway bridge over the M1 near Luton, which he probably built anyway. It is doubtful that Marples was bothered by any adverse publicity considering the number of investigations he was subject to in his lifetime. The point is that there had been an explosion in car ownership and something had to be done to control how they were used and parked.

Drawing lines became something of an obsession with city planners and in 1967 those wonderfully exclusive areas known as 'box junctions' were introduced. This isn't strictly parking however, if you come a standstill in one, then that is something a stationary driver will get done for. Enter them at your peril, or at least when you know that your forward progress is unimpeded. Obviously they made their debut in London and fast forwarding to the present day when there are cameras everywhere has led to accusations that some local authorities have been using them as giant cash machines. That's because their traffic flow plans are designed to funnel drivers into making a mistake that will cost them money.

According to the Highway Code rule 174 'You must not enter the box until your exit road or lane is clear.' However, drivers in cars that are entering the box are obeying a traffic light, which is green, meanwhile the one on the other side is red, so it is no surprise that they are all in a jam. The only exception to the box junction rule is that you can enter it if you planning to turn right. Simply wait until there is a break in oncoming traffic and you are, in theory, in the clear. Meanwhile, there are dozens of capital city based examples, but one box junction in Fulham (Bagley's Lane) earned the council £2.4m in Penalty Charge Notice fines in just 18 months back in 2015.

The thing is local councils could not enforce these moving traffic violations. Under the Traffic Management Act 2004 councils needed to apply for powers to tackle bus lane contraventions, parking and traffic violations. Here was a way of freeing up police time for serious crime and leaving these minor offences to the local authorities who seemed to relish the increased, revenue raising responsibility.

Almost inevitably, wheel clamps were first used by Westminster City Council on the streets of London in May 1983. Simply paying the fine did not mean that a council employee would turn up and unclamp you, the driver had to find the nearest designated payment kiosk, often at a not very near car park or pound. Then return to the car before it could be clamped again. The problem was with these clamps, (originally referred to as the Denver Boot) is that it kept the poorly parked car in place, presumably as an obstruction. The vehicle was effectively held to ransom. Money first, any legal argument would be much later, if at all.

Oddly it took a while for the clamp to make its way into the private sector and that was 1988. Stray onto some private land, which could be anywhere and there could be an unreasonable fee required to release the vehicle. When it came to court, simply missing the disclaimer was no excuse, even if it was partially hidden behind a tree, or wall. According to the Security Industry Authority there were almost 2,000 clamping firms operation in the UK by 2010. It was a very big business and contentious too.

In recent years traffic wardens were downgraded to become the rather less terrifying Civil Enforcement Officers and abolished completely in Scotland from 2014. That has been happening in England and Wales, their work being taken over by the police, just like the old days then.

One thing parking no longer needs is people, or at least not that many of them to administer and enforce it. The first fully automatic payment system came to the Westgate Car Park in Oxford from October 1974. What was novel then, but commonplace now, drivers collected a ticket at the barrier, took the ticket with them while they shopped, paid at a pay station and were issued a ticket to lift the barrier.

Increasingly, the requirement to enter a registration number in a ticket machine, which was then printed, on the physical ticket has meant that it cannot be passed onto another motorist to use the unexpired time. This was a common courtesy, which bonded drivers together and made the world a better and kinder place. Not any more.

Having money in machines was always a huge temptation. Incredibly there were turf wars amongst rival Albanian gangs in London that sought to control of sets of meters that they would access by their own keys. In 2007, Westminster estimated that they were losing up to £120,000 per week in this manner. So time was up for parking meters and they were removed from central London and replaced by something rather more sinister. Cold hard, but neutral cash gave way to chip and pin mobile phone technology instead. The parking bay still has a sign stating the rules and timings, but the extra details required a code to enter and possibly card details, or simply a requirement to text the payment.

The upside would seem to be, not having to worry about having the correct change in your pocket. Paying and renting the space is one thing, which we now do willingly. The trouble is giving up all that extra personal data which can then be shared with others. Be it a big tech company, another local council and/or government departments, never mind the police. Paying for parking indicates that we live in a disciplinary society. Alternatively texting all your personal details suggests that we are actually living in a society of control. That's a bit worrying.

Also, what if you choose to live off grid without a

mobile phone. You can't park, full stop. That seems to be very discriminatory. Elitist even. Bring back the parking meter, at least that didn't judge you.

Certainly The Department for Transport (DfT) are rather excited about all this virtual stuff enabling them to keep a close eye on motorists. They are solidly behind the creation of national parking data standards for local councils and are closely involved in the brave new world of parking without using real money.

Already some councils already have their own competing apps and the DfT are striving for a common parking language, which make it easier for local authorities and private companies to exchange data. The supposed benefit of this will be their ability to help navigate motorists to an available space based on the price, quality of the space, safety of the space and even what on-site services there are. Not necessarily a takeaway, but something useful for an electric vehicle owner, like a charging point. This is crucial in their fight against congested town centres, they say. At the time, Future of Mobility Minister (which seems like quite a nice non-job, where you just speculate about hover buses and Stark Trek Transportation devices, where you could use the phrase 'beam me up Scotty' at least once a day), Michael Ellis, said: "We are on the brink of a revolution for the future of transport, with ground-breaking technologies creating huge opportunities for cleaner, cheaper, safer and more reliable journeys."

This just means they will know where we are, tell us

where to go and presumably, if we have ever been naughty, never find a parking space again.

If you thought double yellow lines were bad, red routes are much, much worse. Even the briefest of incursions lasting seconds results in a heavy fine. They arrived, obviously in London in 1991. They are urban clearways which form a network of major roads and are designed to keep them relatively clear, especially as they carry large volumes of traffic, particularly during rush hour. Drivers can't stop, so in theory there can't be any jams.

However, a single red line means that no vehicle is allowed to stop at any time during the operational hours of the route, which is displayed on a roadside sign. Also there are no U-turns and there will often be lane restrictions.

Whilst we are having a go at bad regulations, there is a lot to moan about when it comes to Controlled Parking Zones. The trouble is that the permitted parking hours for single yellow lines are only signposted at the entrance to the area and not next to the road. That means motorists don't know when they are allowed to park and end up being fined for making a mistake. These zones were never intended to cover large areas, but that seems to be what has ended up happening. Even though they have to comply with quite complicated regulations for legibility and display, motorists often play a guessing game as to when the restrictions apply because there is no nearby sign.

It is possible to go on and on about the frankly baffling and alienating world of local council parking rules. That would overlook the predatory and frankly quite evil, companies that enforce private parking. Basically, Pirates, but not so wholesome or as much fun.

A possibly made up, certainly scary statistic suggests that every five seconds a car owner gets a fine for some parking misdemeanor. Rogue companies came to national attention when they started behaving very badly and often clamping anything that wasn't moving for a second or so and charging a large fortune to release it either from the clamp or a compound.

By 2012, the Government had banned clamping using the Protection of Freedoms Act, but the operators could still send their stiff fines through the post. It was mailed to the vehicle owner rather than the driver. In the past they had to establish who the driver was. At least the fines were limited to £100, but they were so much easier to issue, especially as they could buy driver details from the DVLA for £2.50. Here we have the government colluding with the private sector to screw the motorist.

Anyway, the pirate, sorry private operators who are responsible for public places like hospitals, shopping centres and motorway service areas had their own voluntary codes of conduct. The emphasis here is on voluntary, but they must use proper signage and offer a ten-minute buffer of the allotted stay before resorting to a fine. That does not seem to have reduced the incidents of parking upsets, which stood at around 7m in 2019 alone according to the RAC Foundation.

The upsets don't just seem to be parking on some industrial estate or rough ground where a lack of proper signage dropped the driver in it. Seemingly well run, or should that be oppressively operated, car parks certainly cause problems. Obviously they have ANPR cameras that read registration plates to prove the car and driver were there. That becomes a problem when someone drives in to pick someone up, can't find a space and drives out again, yet still gets a ticket.

Drivers simply overstaying their designated time in a car park is the most common cause of upset which is understandable, from all sides. This scenario does get a different spin every few years. The latest wheeze has been drivers dawdling at petrol stations. It used to be retail parks and train station car parks and in 2019, don't you dare fill up with petrol, do some shopping at the attached Marks and Spencer supermarket, then pump up your tyres and get a coffee. BP and Shell have installed CCTV cameras, so that if a driver overstays their alotted time and have not paid attention to the warning signs, then they will be subject to a £100 fine. It was reported in the Guardian that a company called MET Parking Services working for BP would pursue overstayers and authorise debt collectors when there is no payment.

Another issue has also been that complex ticket machines, which require a registration number misprints it, or logs that number incorrectly.

There are some machines, which require users to press a button to trigger an evening rate, but they need to be in the know to activate it, resulting in the incorrect fee being paid and a fine issued.

If the machines don't work, what was the driver supposed to do? Just ignoring the issue and getting on with the business of parking results in a fine.

Something had to be done and that was the new Parking (Code of Practice) Act 2019, to bring in greater regulation. Drivers will be able to easily challenge unfair parking tickets through a new independent appeals service, and all private parking operators will also have to follow a new industry backed Code of Practice. If private parking firms break it then they could be barred from asking for motorists' information from the DVLA to enforce tickets. The new independent appeals service will also give drivers greater support to challenge unjustified parking tickets.

However, just when you thought it was safe to find a parking in the real world and anyway, you have your own designated space outside your office. However, the authorities have already thought of that. Here are three truly terrifying words: workplace, parking, levy.

Nottingham was the first council to introduce the scheme in 2012 and it seems to have had almost no impact on congestion. In 2018 it raised over £9m in a single year. The money raised is supposed to be spent on sustainable transport projects. Under their scheme employers who provide more than 10 parking spaces for their staff pay about £415 every year to the city council for each space, with the charge increasing each year in line with inflation. There are some important exceptions, as there is no charge for motorbikes, customer vehicles, fleet vehicles that are parked at an employer's premises, but are not used to travel to and from work, or vehicles that are used to deliver and collect goods.

That makes sense and so does the fact that staff parking at hospitals and other NHS premises are exempt from the charge, as are disabled parking spaces and front line emergency services such as the ambulance, police and fire services.

The trouble is other local authorities are revving up to create their own schemes. That means there is a lot more of this to look forward to. It will be justified on the grounds of cutting congestion, improving the air quality and obviously charging more tax. The thing is that company owners already pay business rates for having a car park. The simple fact is that extra costs are passed onto workers and consumers. If you really want to improve the journey to work, fix potholes and make sure that the public transport alternatives are reliable.

Essentially charging to park at work in the city with buses every 5 minutes? Hard to argue too much with that. Charging to park at work on an industrial estate in the middle of nowhere with minimal public transport? That is unfair.

Surely parking should not be either difficult or oppressive, just allow the ordinary motorist to go about their business, be it work, shopping, or just having a laugh.

Roads do need to be managed so that they are safer and less congested. The world as we know it might just come to a halt if there was no rules, structure, or yellow lines. Space though comes at a premium. Organising and administrating that space, even painting the lines has to be paid for. It is also a system for control. The trouble is that parking is a business. An industry. This has created a massive self-sustaining bureaucracy that succeeds in pissing people off because it is not fair.

Parking is all about packing as many cars into the smallest possible area. It is significant that there are, apart from disabled parking, no set sizes for slots. It is incredible that as vehicles have grown, the places to put them most certainly haven't. According to the British Parking Association the norm for off street bays is 2.4 metres wide by 4.8 metres long. However, it could be anything. For on street bays there is The Traffic Signs Regulations and General Directions 2016 but there is no minimum height or width, just about big enough for a modern car. Just a note that Disabled Persons Transport Advisory Committee has minimum dimensions that are prescribed for bays reserved for disabled badge holders. These must be a minimum of 6.6 m long, 2.7 m wide, or

3 m wide.

The highways authority arguably has one simple purpose and that is to provide an uninterrupted route from A to B. There is no obligation to provide or even surface any parking area. They do roads. Not stopping.

Town centres may be dying off and it is easy to blame the Internet, but short sighted planners allowing out of town hypermarkets where parking is free, certainly for a one to two hour period in most cases, has sucked the life out of them. Then overregulating parking, or simply free ticketed parking in these towns can lead to drivers and that means shoppers, giving them a miss.

It is worth saying that there are too many people with too many cars in certain areas. Officially the population is 60 odd million. The government actually has no idea how many people live in Britain, which is a scandal. The National Census is becoming less reliable every time they are conducted. If net migration is running at 300,000 annually then having to cope with a new City every year means there is less room and resources not just for drivers, but everyone. Clamp down on all that, raise driving standards and just maybe the whole space programme could start to sort itself out.

*It was always cheaper to ride on the 'hard shoulder'*

## 7. The Price is Wrong - *Pay As You Drive.*

Pay as you go. Sounds very reasonable. We are all used to this concept because of phone contracts and other modern ways of settling up for things and the time seems to be right, for all sorts of mainly political reasons, to apply it universally across the UK to motorists.

Officially in the UK it has been called 'Road pricing'. This is a system of charging drivers to reflect car usage, with regular vehicle users paying more than some old dear who just uses her Honda Jazz once a week to buy cat food. Not only that, these schemes are designed to spread out actual road use throughout the day. So it would be expensive at rush hour and dead cheap in the middle of the night. Drivers would travel when they could best afford it and congestion may not be avoided, but certainly reduced. It is of course very fair to the old dear in the Jazz, whereas Gavin the bastard in the Jaguar who needs to open his carpet warehouse pays more. The thing is though Darren, his averagely paid carpet warehouse manager needs to be there at the same time but there is no bus which goes anywhere near the industrial estate where Carpetz Carpetz Carpetz! Is based. Like so many taxes, in the real world they are not fair.

Although Road Pricing is not yet in the UK, charging to use a particular patch of road or crossing has been around for some time. They were called Toll Roads and way back in British history there were Turnpike Roads, named after the frame of pikes that could be turned to allow horses through. By the late 1600s British roads were in an appalling state, having had no major investment since Roman times. Yes what did they actually do for us? Build roads, then when they went, no one else could be bothered. In 1656 the boroughs of Hertfordshire, Huntington and Cambridgeshire faced bankruptcy with the cost of maintaining the Great North Road into London, so they lobbied parliament for the right to raise tolls to pay for the upkeep. The Turnpike was born and the system spread throughout the 18th Century, until at their height in the early 1800s, about 1,000 trusts controlled 18,000 miles of road in England and Wales.

Not everyone was a fan and there were protests where it threatened farmer's livelihoods by restricting their movement. Mostly though the Turnpike was a focus for general discontent about a lot of other things. Turnpike trusts were eventually rendered obsolete and often bankrupt by the railways and canals that the roads had effectively helped to build. It was The Local Government Act of 1888 that handed responsibility for roads back to councils. Roads wouldn't have to get better again until the motorist fully arrived and along with road tax. For the time being though the British public were very much out of love with the whole Turnpike concept. Presumably the authorities think we have forgotten all about that.

Before we get to road charging a few words about the tolls which already exist in Britain. There are not many, just 23 tolls and some 18 of those take you across a river. The best-known or most notorious ones includes the M6 Toll just outside of Birmingham, which takes traffic around the snarl up on the actual M6 and provides a handy shortcut. Then there is the unavoidable Dartford River Crossing, whether motorists go over the bridge or through the tunnel, they should not be paying a penny.

Under the original agreement when the bridge was built, tolling was supposed to stop once it had paid for itself. Indeed, back in February 1999, the government announced that the Dartford Crossing would be free of toll charges by the end of 2003.

Under the original agreement with the Ministry of Transport, both Kent and Essex County Councils got government approval to charge tolls in 1960. The toll was originally two shillings and sixpence. Later the Dartford Tunnel Act 1967 gave the County Councils authority to change the tolls, and in December 1977, the toll was raised from 25p to 35p for cars, 40p to 55p for 2 axle goods vehicles, and 60p to 85p for HGVs. Obviously the charge kept on going up by pennies over the years, but hey, once it was all paid for, then drivers could spend that money on something else. Maybe a decent sarnie at lunchtime, or overtaxed petrol to put in the vehicles they are using.

However, it came to light 2001 that the government

had decided that the original agreement needed some modification. That income stream was too much of a good thing to give up. The tolls would stay, hardly surprising as this gateway between, the garden of England (Kent) and home of longest pleasure pier in the world at Southend on Sea (Essex) is the busiest stretch of road in the UK. The Transport Act 2000 allowed the charges to continue and they were effective from 2003. That original promise to the motorist was broken.

The trust that operated toll said additional income would be used to fund local transport and the Department for Transport confirming this. Motorists remained remarkably compliant, probably because they still needed to get to work. The Department was mindful that removing the tolls would lead to unprecedented levels of traffic. That's right, they were doing the existing commuters a favour.

Then in 2014 the tollbooths were removed. Not so that drivers could travel for free, but precisely because it was possible to make payment more efficient and potentially levy even more fines, never mind allowing foreign registered vehicles to dodge it completely.

Between 2015 and 2016, the first complete accounting year during which the barrier less system was in operation, £76.7m was paid by the discounted 'Dart Charge' account holders - individuals and organisations who regularly use the Dartford crossing. An additional £31.3m was paid by those without Dart Charge accounts, either online, by phone, mail or Payzone.

However, enforcement revenue generated through fines for non-payers reached a rather high £53.1m. Significantly more than 2 million cases were passed to the European Debt Recovery Agent to try and track down foreign registered toll dodgers. Trouble is compared to British motorists who are legally obliged to be registered with the DVLA, means tracing foreign registered toll dodgers difficult. Again, the Brit motorist will always pay a disproportionately large amount of the fines.

All this makes anyone not already living in Scotland, keen to move there. Never mind the lovely inhabitants and stunning views of Lochs and the Highlands, the best part of being north of the border is that they abolished toll roads in 2008.

Nevertheless, road charging is very much back on the agenda. As politicians wonder where their tax revenue will come from when electric vehicles take over the roads. Back in 2005 Sir Rod Eddington reported to the Labour government that road pricing could help tackle congestion.

A government study recommended a sliding scale of charges, rising from 2p a mile to a maximum of £1.34 a mile for the most congested roads. That was just a finger in the air, testing which way the wind was blowing, guess. Things actually progressed as far as some local authorities proposing their own schemes. That resulted in more than £14 million being split amongst ten areas, including Bristol, Cambridge, Tyne and Wear and the West Midlands. In 2006 it spread to Reading, Norwich and Nottingham, Leicester and Derby.

The good news is that motorists didn't just sit back and take it. 1.7 million people signed a Downing Street petition in 2007. To highlight just one of the paragraphs which succinctly sums up how drivers felt, "The idea of tracking every vehicle at all times is sinister and wrong. Road pricing is already here with the high level of taxation on fuel. The more you travel - the more tax you pay. It will be an unfair tax on those who live apart from families and poorer people who will not be able to afford the high monthly costs."

The government denied that it would turn into a stealth tax, or track driver's journeys and justified exploring road charging, purely on the grounds of congestion.

'Public attitudes towards road congestion' was produced in 2010 for the Department of Transport. Apparently more than half of adults agreed that the system should change and be based on use. No one got to find out of this was indeed true as the incoming Coalition Government did not act on these findings.

The same year an RAC Foundation report, 'Governing and Paying for England's Roads', concluded that some form of 'pay as you go' system would need to be introduced to deal with congestion. The idea of paying to drive has never entirely gone away and will be revived on environmental or revenue raising grounds eventually.

Nine years later in 2019 and Members of Parliament wanted to start a brand new road pricing debate as well as introducing workplace parking levies. This was as a result of the cross-party Transport Select Committee, which will be starting a formal inquiry in 2020.

Potentially motorists could be faced with pay-per-mile road charging and other taxes designed to plug a £40 billion shortfall caused by electric cars, which are exempt from road tax and generate no fuel duty revenue. Interestingly the Committee said that “road pricing does not only mean tolls”, what it may mean then is congestion and low emission zones which you can read a bit about elsewhere in Demotorized.

We might as well speculate over what might happen in the near future. There could well be pay-per-mile charging as a result of GPS data loggers, which will be a standard fitting on new cars from 2022. That means that the UK’s ANPR camera network could be used to charge drivers based on the length of their journeys.

Nothing really changes, apart from the technology, as the motorist will always be compelled to pay for the privilege of driving.

*ULEZ Compliant*

*Xhibit A*

*Vehicle not entering ULEZ*

## 8. Zombie Zones - *The shocking price of entry.*

Nuclear free zones. Remember those? If you were lucky enough to live in the people's republic of Brent (that's in London) back in the 1980s you were safe. Not only were there no nuclear weapons in the borough, it wouldn't be on the list of targets for The Union of Soviet Socialist Republics. Spare a thought for the poor residents who lived next door in Barnet. They could have been vapourised at any time.

Ultra Low Emission Zones ULEZ really are that daft. In theory it means localised cleaner air, but often the problem just gets shifted somewhere else, or if the wind blows in the wrong direction then it will come back at you, no matter how free, or low emission the zone is.

Right back in the beginning though, exclusion zones didn't have very much to do with air quality, which is something you can't actually see. What you could touch, smell and see though, was congestion.

As usual, the place to conduct any social or political experiment is London. It is full of people with very little to unify them these days and divided up into lots of Brents and Barnets. Very few of the inhabitants will object or stand up for themselves anymore. It is sadly a fractured and divided society, so perfect to impose nonsensical schemes upon. London is also rich enough, certainly in the centre, not care about the cost of anything going up by a few quid a day, week, month or year. They are rich enough to cough up money to go on living and working where they are. This is also where many politicians live. Such measures won't affect them or are at least are claimable on expenses. It is also a conveniently close so they can virtue signal about it if they need to. Welcome to London's Congestion Zone. Established in 2003 and designed to magically banish all pointless journeys and vehicles by the imposition of a charge.

The Zone has expanded a bit over the years and covers most of central London with Euston Road in the north, Commercial Street in the east, Vauxhall and Chelsea Embankment in the south and Harrow Road and Earls Court in the west as the boundaries. The scheme is monitored by cameras on roads across the congestion charge zone, which read car number plates and cross-references them against a register of cars that have paid the charge.

The mayor of London at the time, Ken Livingstone, introduced the congestion charge at a bargain £5 a day in February 2003 within a designated charging zone. The timings have changed slightly and it currently operates between 07:00 and 18:00 on Monday to Friday, excluding public holidays. Not everyone paid and some got discounts, this included people with disabilities, residents living within the congestion zone, emergency services and breakdown recovery vehicles, taxis, and drivers of alternative fuel vehicles. More on that elsewhere.

Several embassies don't play the congestion charge game and use the old 'diplomatic immunity' get out clause, last seen in Lethal Weapon 2. They believe that Vienna Convention fine print means they can simply ignore local taxes, never mind parking tickets. This is an area of interest in itself. The Metro newspaper in 2019 discovered that the worst offenders were America (£12.5m), Japan (£8.5m), Nigeria (£7m) and India (£6m). At the very least that gives you some idea of just how much income the scheme generates, even if there are some dodging it, to the tune of £116m.

As well as unblocking London that charge was also designed to collect net revenues to improve public transport facilities in London, as, by law, the proceeds raised by the congestion charge were then transferred to the London public transport purse. The stated aims included reducing the length of journeys within the congestion zone, improved bus services and to encourage motorists to use public transport instead of their cars.

Whilst all that important work was going on, the charge was increased to £8 a day from July 2005 and to £10 from January 2011. There are several methods of payment, including telephone, text message, post, online or in a corner shop. Failure to pay risked a penalty charge notice of £120.

Operation of the scheme was outsourced to Capita RAS and despite getting an extra £31 million they were not exactly brilliant at it. In October 2003, the firm was fined £1 million for its shoddy performance. They lost the contract to IBM in 2009 and won it back in 2014 and there is talk it will be administered in house by Transport for London at some point.

The congestion charge is like so many taxes, regressive. So the poorer motorists are charged the same levy as richer car owners. There is no exemption for those regarded as key workers, be they NHS, Fire Service or Police, or just someone on the minimum wage who is obliged to work in the congestion zone at chargeable times.

Any business situated within the zone has to pass on that charge. It isn't going to absorb any costs unless they want to lose money, it has to charge more for goods or services. The charge may also put off potential customers from using the company's goods or services. Everything in the zones simply becomes more expensive and some politicians might regard that as a price worth paying. There were stats on this when in 2003 the London Chamber of Commerce found 79 per cent of just over 200 shops reported a fall in takings and 42 per cent said the congestion charge was to blame. Taking that survey a couple of years later a massive 84 per cent of those shops reported a fall in takings, with 62 per cent attributing this to the charge.

The charge could have become seriously expensive if it wasn't for an election. However, the lower emission zone was making its debut when in 2008 Ken Livingstone suggested the charge for the most polluting vehicles should be £25. The arrival of Boris Johnson, as a Mayor put an end to that. He also stopped it getting bigger with a Western extension and prevented the low emission scheme being extended to vans, for the time being at least.

One unintended consequence of the charge wasn't any actual reduction in congestion. In 2007 Transport for London estimated that there were an extra 1000 deaths as a result of poor air quality and a further 1000 hospital admissions, but these had all the appearance of made up statistics unsubstantiated by actual medical facts.

However, King's College, London and the London School of Hygiene and Tropical Medicine published its findings in everyone's favourite magazine, Occupational and Environmental Medicine in 2008. It showed that Londoners were now benefitting from a significant reduction in nasty old nitrogen dioxide. Scientists from two London colleges calculated that since 2003, 1,888 extra years of life had been saved among the city's then seven million residents. Here was all the evidence they needed to make a motorists life harder and more expensive.

Anyway, back to The Low Emission Zone LEZ, which came into operation from February 2008. It all starts to get a bit complicated from now on. Essentially the emission standards are tougher of course and vehicles registered after October 2001 can go into the zone with an engine that complies with Euro 3 emissions standards. Older diesel buses were culled from 2012 and other buses were converted to Euro 3 or 4 standards. Without getting lost in the rules and regs effectively what Londoners had to deal with here was another revenue raising ruse.

In 2017, London Mayor Sadiq Khan introduced an additional £10 toxicity charge, or the T-charge for older pre 2006 diesel and petrol cars in response to some high recorded levels of pollution. That bumped an extra £11.50 on top of the existing charge. It sort of begged the question that if things are so bad, just ban those vehicles outright. Why take the money and still cough?

Never mind because there was the sound of something else hovering overhead. What's that being dropped into place around London from a great height? Well it must be the brand new Ultra Low Emission Zone ULEZ, which arrived in 2019. What this meant was any diesel not conforming to Euro 6 emission standards and any petrol engine not conforming to Euro 4 emission standards was banned. Petrol cars that meet the ULEZ standards are generally those registered with the DVLA after 2005, although cars that meet the standards have been available since 2001 and you can find old Honda Civics, which currently comply. When it comes to diesel cars that meet the standards, these are generally registered with the DVLA after September 2015. For owners of vans, then a rare petrol one must be Euro 4 (from 2006) and a diesel: Euro 6 (from 2016). Many owners have been caught by the manufacturer as sometimes the vehicles had been registered a long time after they were built. The V5 registration document is always going to be the definitive proof as whether the vehicle qualifies, or not.

For those in the heavy goods vehicle business their lorries would have to be Euro VI standard. Here though is the anomaly, as all Motorcycles and mopeds, which were, previously not subject to the charge must meet Euro 3 emissions standards. Usually it will mean bikes registered with the DVLA after July 2007.

Those after a true congestion buster, should be on the side of the two wheeled. Bikes, whether they are called motorcycles, scooters or mopeds are easy to park. The congestion issue is fractionalised because they are so small, but pollution seems to be their downfall. It hasn't looked especially good over the years as on the purely emissions aspect.

Less CO2 but more hydrocarbons, particulates, and nitrogen oxides. Most vehicle pollution is caused from burning fuel. If a scooter gets 100 mpg and a car a very poor 20 mpg, the scooter is producing 1/5th the pollution from fuel as the car. Looking at simple efficiency ratings then a bike is usually less than half that of a typical car. But if a car is crawling along in traffic with the air conditioning on while the bike keeps moving and completes the journey in a fraction of the time then the bike starts to look much more efficient. There is also the point that many smaller scooters use two stroke engines. Two stroke engines are lubricated by a fuel additive that is burned together with the fuel. Two strokes are highly polluting compared to four stroke petrol engines in cars. Bikes though are just less of everything and if the congestion charge was ever about reducing traffic jams then more motorcycles would make sense.

It doesn't help that many of the simpler scooters lack much emissions gear and have simple exhaust systems. There is though just the one cylinder in most cases pumping out the bad stuff rather than four or more. However, what is never seems to be factored into its environmental impact are the whole life costs. Building one is less resource intensive, plus the mpg shames all four wheeled vehicles as the fuel economy is more than double a car. As a result ULEZ condemns perfectly good motorcycles, which have become almost worthless and are probably scrapped if they can't be used elsewhere.

One issue is that a whole category of interesting vehicles is not just being excluded, but like the bikes, could end up being scrapped anyway. Between modern environmentally correct cars and 40 year old plus exempt is a swathe of great future classics and important ones that can't beat the zone, so are sold on or got rid of. They won't be practical propositions, so our motoring history becomes poorer as a result. There is a very good detailed blog on this by journalist Mark McArthur Christie. Reference at the back of this book. The point is that there is no distinction made between a well maintained 35 year old classic Mercedes, or a 25 year old diesel Renault belching out smoke.

For those who want to avoid the charge altogether, should strap an ULEZ compliant moped to their bottom and spend three years learning the knowledge, then buy a black cab for a large fortune. Otherwise it is the usual exempt suspects including residents, the disabled and historic vehicles, so as mentioned, characterful 40+ year olds are allowed.

When it comes to freeing up the streets the single worst thing that has been done was making an exception for taxis. Not the proper black ones, the hybrid ones which clog the central London area. Private cars have disappeared, it is mainly commercial vehicles like vans and Ubers. (Read about Uber elsewhere in Demotorized). The congestion zone charge is now applied to them and it will inevitably affect fares, but it is not enough to make those who have made a career picking up passengers want to give up. Meanwhile taxi app users would be loath to give up the convenience.

Taking a cab might make life a lot easier. Otherwise there is an online ULEZ vehicle checker to find out the payment levels. When Demotorized was published, the London charge was £12.50 for everything, including bikes, up to and including 3.5 tonne vehicles and £100 for anything heavier. All of this may not scare you much because you never go to London and now never have any intention of driving there at all. Hold on though, here's a news flash, ULEZ is coming to a built up area near you.

Aberdeen, Bath, Birmingham, Bristol, Cambridge, Cardiff, Derby, Dundee, Edinburgh, Fareham, Glasgow, Leeds, Manchester Newcastle, Oxford, Reading Sheffield, Slough, Warrington, York, to name just a few. Even at the planning stage when implementation might be a year or two away, there will be controversy. Bristol's, no private diesel ban has put the cool clean air cats, amongst the dirty old pigeons.

That seemed like a rather excellent place to cap the chapter on a nice turn of phrase, but there was more. Especially when it came to double standards.

Obviously, the vast majority of all the 320 councils (England, Scotland, Wales and Northern Ireland) who may be imposing ULEZ schemes run diesel vehicles. Auto Express did the investigation in December 2019 and confirmed that most (91.6 per cent ) of the 66,617 vehicles operated by UK councils run on diesel. In all, 62 council fleets consist entirely of diesel vehicles.

Auto Express found that the worst offenders were: London (89 per cent), Newcastle (93 per cent), Leeds (86.5 per cent), Manchester (97.8 per cent), Sheffield (92.2 per cent), Oxford (80.3 per cent), Bristol (81.5 per cent), Bath & North East Somerset (96 per cent), Aberdeen (95 per cent), Dundee (81.8 per cent), Glasgow (98.2 per cent), Edinburgh (94 per cent).

Amongst the two most hypocritical councils was Bristol with their own fleet of 453 vehicles including 369 diesels. Not only that, they were buying 64 new diesel vans. Meanwhile in London drivers of pre-Euro 6 diesel vehicles were charged £12.50 to enter the city centre ultra-low emission zone. Yet nine out of ten (89 per cent) of the 4,844 vehicles operated by London councils ran on diesel. At least 724 (15 per cent) were pre-Euro 6 diesels that are not ULEZ-compliant.

There you have it, the local councils setting a clean air agenda, but not practicing what they charge motorists for preaching. Council vehicle fleets will change eventually, but this was not a good start.

Drive what we tell you to drive and pay dearly for it.

Please vote for us next time.

Remember that.

*Prototype Victorian SUV*

## 9. Carmakers vs. Customers - *Why do they hate their buyers?*

Most car manufacturers are happy to be friendly with their customers. Not to go on holiday together or buy reciprocal secret Santas or anything, just get along with each other. The trouble happens when the vehicle becomes problematic, a bit rubbish, or is more dangerous than it really should be.

When it comes to making changes to what seems to be a perfectly good vehicle, just for the sake of it, well, that's called marketing. American journalist and author, Vance Packard explained built in obsolescence back in 1957 in his seminal work: The Waste Makers. It is all about marketing and snobbery, so everyone knows you drive the very latest, or just own the most up to date thing and are putting your neighbours or work colleagues in their place with a good old fashioned example of one upmanship. Corners though, could be cut.

In Britain things were not too bad, it didn't seem as if the carmakers wanted to actively kill us. In the United States though, money came some way ahead of customer satisfaction and safety. Also, Americans certainly seemed to like writing books about how much they hated their cars.

Unsafe at Any Speed: The Designed-In Dangers of the American Automobile by Ralph Nader and published in 1965 made it clear that all was not well, particularly with the Chevrolet Corvair. It is worth pointing out though that Nader certainly had an agenda and didn't seem to like any cars, but he did have a thing about the Corvair, which he reckoned was a "one-car accident."

Did he have a point? Essentially, this was a very un-American design. Instead of a big V8 engine up front driving the rear wheels, there was a flat air-cooled six-cylinder engine plumbed into the boot, driving the rear wheels. It also had an unconventional swing-axle suspension system, which combined with incorrect tyre pressures caused the handling to be more than wayward. He may have been correct but the fundamental problem seemed to be the rear engine and weight distribution that most American drivers were just not used to. Some owners didn't have a problem, unless they drove it in a sprightly manner without allowing for the change in weight distribution. If they did have an incident, the other terrifying thing was that the hinges holding the bonnet on were not very good and they could fly off in a dangerously decapitating manner.

The first chapter is all about the Corvair and the rest of the book went on about other makes and models. The trouble is that the first chapter took all the attention and was so boring that no one bothered to read any further. Certainly rear engined Volkswagen Beetles of the same era seemed to be just as lethal at flipping over. Also, Porsche 911s had a very similar design and handling issues for drivers who were not paying attention. However, from ’65 onwards the Corvair had independent suspension and that seemed to make a big difference.

Oddly enough the National Highway Traffic Safety Administration in 1972 three years after the model was discontinued reckoned it wasn’t so bad. They said that the early model which had the issues, “compared favorably with the other contemporary vehicles used in the NHTSA Input Response Tests. The handling and stability performance of the 1960-1963 Corvair does not result in an abnormal potential for loss of control or rollover and it is at least as good as the performance of some contemporary vehicles both foreign and domestic.”

Nader was a politician and a lawyer, a lethally bad combination of professions there, as well as running for President four times. He did seem to have a point though that the design had not been thought through, but it may not have been as bad as what came later. His activism led to the National Traffic and Motor Vehicle Safety Act in 1966, which meant there were minimum safety standards for all vehicles sold in the States.

It is worth staying in the States for a bit longer because that Act didn't stop what happened next, which is the single worst example of manufacturer sanctioned customer casualty in automotive history.

There isn’t anything funny to say about the Ford Pinto, also badged as a Mercury Bobcat, an unremarkable small saloon and hatchback (effectively an American Ford Escort) that seemed to have a propensity for catching on fire when struck from the rear. The victims burned to death rather than suffering fatal collision injuries. The body buckled and doors jammed but the problem was that fuel escaped and ignited because of a poorly designed strap on the tank.

In all, 27 people were killed and many more seriously injured. Not very many cases reached trial, when they did the awards were huge, usually substantially reduced on appeal, but it really did boil down to money and having very good lawyers.

First of all there were several potential fixes, which included a rubber fuel tank liner and simply adding a steel plate to the rear of the car. All relatively cheap to do at $5-$10 although a piece of moulded plastic to stop anything like suspension bolt penetrating the tank was less than a dollar. The reality was that redesigning the car to make it less likely to catch on fire would cost well over $100m. Meanwhile, settling with those who had been injured or died was less than half that.

According to the legal eagles, a human life in mid '70s America was worth around $200,000 and a serious burn $67,000. There was a lot of public pressure by now and an official investigation by the National Highway Traffic Safety Administration's Office of Defect Investigations and consumer expose in the Mother Jones Magazine, called "Pinto Madness". Not only that in a Californian court there was a damages award of $125million in

Grimshaw vs. Ford, reduced to $3.5million at the main trial.

Incredibly it took until June 1978 for Ford to issue a recall for 1.5 million 1971-76 Pintos and 30,000 Bobcats. A new, longer fuel-tank filler neck and a protective plastic barrier between the rear wheels and the tank was the manufacturer fix. Ford knew they were culpable, but they relied on their army of lawyers to do the wrong thing. We know this because Lee Iacocca told us in his autobiography. Developing the Pinto, as smaller, cheaper, more fuel-efficient American made car to fight off the European and Japanese imports came at a price. In 'Talking Straight' he wrote, "Clamming up is what we did at Ford in the late '70s when we were bombarded with suits over the Pinto, which was involved in a lot of gas tank fires. The suits might have bankrupted the company, so we kept our mouths shut for fear of saying anything that just one jury might have construed as an admission of guilt. Winning in court was our top priority; nothing else mattered. And of course, our silence added to all the suspicions people had about us and the car."

There you have it, a good job we had nothing as lethal in Britain. Indeed, Ford had the good old Capri, which was actually exported to America and sold by Ford subsidiary Mercury. It was a similar size to the Pinto, but the European's fuel tank was out of the way of the rear bumper, so less chance of it catching alight. Clearly building a car that would not explode over a relatively modest rear end incident was not difficult to do.

In Britain there was a long tradition of making bad cars and then leaving customers to sort it all out. With a year's guarantee the car didn't have to last many months being patched up at one of their garages before the twelve months ran out. When it comes to contempt for the customer, we really don't have enough space to cover the rubbish that the British car buyer was expected to drive.

Instead let's take a very quick look at the sort of car that the car industry thought was good enough for Johnny Foreigner to endure in the post war export or die environment of 1948.

The Standard Vanguard, which was bravely billed as 'Made in Britain – designed for the World.' The bodywork with no separate boot was similar in some ways to the Volkswagen Beetle, but it was bigger and had two more doors. Its body was attached to a separate chassis, which had all the important oily bits like the engine, gearbox and prop shaft that took the power from the front to the rear wheels. The gearbox was a three-speeder and it had something called synchromesh on the first gear, which meant that there was no nasty grinding noises anymore when drivers engaged that gear. There was a reasonably economical 2-litre engine and it cost £544. Between 1948 and 1952 185,000 found buyers. So it must have been brilliant.

Standard was staffed by engineers and designers who in

most cases had not been outside of Britain. The prototype Vanguards had only been taken as far as Wales. There the steep hills and farm tracks were the nearest they could get to replicating overseas terrain. Given the state of Europe and limited time available, it was understandable that they didn't stray too far. Maybe they should have popped across the Channel to Belgium to drive up and down on that weird pave surface that used to be so common. Who would have thought that a brick weave surface could have done so much harm? Certainly not Standard who dispatched their Service Manager for Scotland, Freddie Toop (on the basis that he was one of the few employees with a passport) to find out.

What he discovered was that the suspension and whole structure of the Vanguard was being shaken to bits. There were cracks in the metal, which had occurred after just a few thousand miles. The suspension parts called shock absorbers had absorbed all the shocks they possibly could and then broken. As a stop gap the Belgians topped up the fluid in the absorbers with extra fluid, which only succeeded in making them solid. So when the Vanguard hit a bump eventually it would fracture the bolts holding it to the chassis half of the car and they would eventually punch their way through the front wings. It must have been a spectacular if a frankly terrifying experience. Clearly the Vanguard wasn't suited to anywhere more challenging than rural Wales. It stayed in production until 1963 in various restyled and reworked ways with a better and bigger engine, but obviously it was all too late and sales in the UK never mind anywhere else in the rutted, paved or potholed world, were marginal.

As all the small car companies came together first as British Motor Corporation BMC, then becoming British Leyland, no lessons were learned at all. The bigger the company the worse the problems and the harder they are to sort out, even if they want to. The very last incarnation of the British Car Industry was just as bad as the BL days. Indeed, the MG Rover Directors did their best to con everyone including their loyal and long suffering customers, a private company that was run into the ground for the benefit of the nominal owners.

On the whole though, state owned car companies don't have a good record for dealing with their captive audience of customers. Indeed they make them wait. Take the Volkswagen Beetle. In 1938 a production plant – the Volkswagenwerke or 'people's car factory" was built on a soggy plain at Wolfsburg and modeled on Ford's own factory at Rouge River in Detroit. Indeed, Hitler admired the Ford organisation and certainly Henry Ford leaned a little bit to the right himself. Ford sent Hitler a cheque for fifty thousand dollars on his Birthday, when possibly a card would have been sufficient. Unlike Ford, Hitler didn't rely on private enterprise to fund his factory, this was going to be very fascist undertaking using money confiscated from the trade unions and built using Italian construction workers on loan from nearby dictator, Mussolini. The new car which would cost 999 Reichsmarks was to be financed by a savings plan, a sort of Green Shield Stamps scheme (you collected stamps gained from petrol purchases and elsewhere, licked them, stuck them in a book and when you several thousand green stamps you could exchange them for a steam iron in a big shop: that business eventually became Argos), but rather more red and it turns out something of a scam. However, 270,000 had subscribed by 1939 and despite collecting the red stamps which featured the white outline of the KdF (Beetle), faithfully sticking them into savings books, and paying five Reichmarks a week to the German Labour Front none of the 'people' actually got to own one.

That was National Socialism for you, the German Democratic Republic that occupied half of the country after the war wasn't much better. At least they had an actual vehicle, which was sort of thrown together.

It is time then to pick a much easier target and remember that there were in fact two distinct German car industries and one company summed that situation perfectly. Trabant.

The Trabant had a steel chassis and frame that was made on huge presses which stamped out the basic shapes in a poisonous and deafening atmosphere. The bodywork was thought to be fibreglass, but it was hardly as sophisticated as that. Duroplast sounded good but the ingredients shredded plastic, brown paper and cotton waste, all soaked in resin. Overall these was not ideal working conditions and if the workers themselves survived all that and some didn't, they would then have to wait up 14 years to get to the front of queue. For what was a rubbish car.

British Leyland was not much better at customer service. Happy to have waiting lists for Jaguars and Rovers which when they arrived were not worth the wait. The build quality with terrible paintwork, poor panel gaps and appalling assembly, was never going to be sorted out by an indifferent dealer network. Please read 'The British Car Industry Our Part in Its Downfall' for lots more details.

There are so many easy targets, such as the terrible handling of the hurriedly introduced Morris Marina. The terminally overheating Triumph Stag. The dumpy and unreliable Austin Allegro. However, that was back in the old days, because surely things have improved since the grim old 1970s?

Land Rover unreliability is almost legendary, although it does not seem to be stopping buyers paying out serious money for what is a very expensive luxury off roader, which should not break down. There are plenty of reliability surveys, which show considerable customer dissatisfaction with Land Rovers. However, it is always better to look at results based on actual cars that have broken down, have been repaired and subject to warranty claims. In 2019 Motoreasy analysed vehicles that were between 3 and 15 years old and covered an incredible, 40,000 years' worth of warranty data.

It turns out that Land Rover is not only expensive to remedy - with average repair costs cheaper than only Porsche and Lexus, but were also second from bottom in terms of their frequency of failure.

Almost a third of Land Rover owners - 32 per cent - experience a fault each year, putting the British firm on a par with Chevrolet in the breakdown stakes. According to Duncan McClure Fisher, Founder and CEO of MotorEasy: "Land Rover is an iconic British brand, but our rankings show it has some work to do on the reliability front – they keep telling us their latest cars are more reliable but our stats show that if you own a Land Rover you must have warranty protection."

So that buyers don't fall victim to a car that was as dangerous as the Ford Pinto, or as sloppily thrown together as a Morris Marina there is a now a much more sophisticated recall system worldwide.

Please turn over...

*Crash and indeed, burn.*

## 10. Total Recall and Total sCrap - *Fixing bits and new cars for old.*

There are times when the authorities will come for your car. Sometimes it will be an official approach, otherwise it may well be a very simple cash bribe. Let's start with the upfront, polite way when they request your vehicle's appearance at your most local dealer so that they can fix it.

What the recall system proves is that car manufacturers never seem to learn their lesson. Certainly cars are fabulously complicated pieces of machinery, maybe too complicated. That's why manufacturers are obliged to sort out defects by working with the Government to find your address and then send you a letter.

Officially a recall happens when a car manufacturer wants to check one of their brilliant products, for safety reasons. It won't mean that it is now unsafe and undriveable, unless the manufacturer says otherwise of course.

If it is really bad, serious and scary then motorists may find out on the national news that their car is being recalled. Usually a letter is sent to registered owners detailing why it is being recalled, what the owner needs to do and who to contact. It will refer directly to the Vehicle Identification Number (VIN) on the car.

If a car recall is ignored and there is no legal obligation to follow it, car owners are required to keep a vehicle in a safe and roadworthy condition. Plus if there was an accident as a result of the recall fault, it may be difficult to make an insurance claim. Drivers also have the option to report defects independently to the Driver and Vehicle Standards Agency.

Oddly enough dealers quite like recalls. Firstly they get paid to fix the cars and it introduces them to someone who most likely is not a customer. There are also some unofficial recalls, which are called Technical Service Bulletins issued by manufacturers. Usually it is a minor component that needs to be adjusted or replaced when a vehicle is next serviced. These bulletins are rarely made public. It is under the radar then and designed to make cars perform better, but just not serious enough to be a proper recall. Well, that's what they say anyway.

Certainly, the global nature of building cars these days means that problems may not actually originate with the manufacturer. Increasingly the really big problems have been expensively outsourced. There are many examples and the most high profile worldwide mechanical muck up has come courtesy of Japanese company Takata and their exploding airbags. Rather than simply inflating and possibly giving the front seat driver and passenger nose jobs, theirs could possibly result in a full scale realignment of all their facial features. The official statement said, "Over time and in the event of an accident, the airbags may inflate with such force that parts of the airbag unit could shatter firing metal fragments into the cabin." Yikes.

Around the world these defective airbags could be found in BMWs, Hondas, Chryslers, Lexusessesss, Jeeps, Subarus, Mazdas, Mitsubishis, Nissans, Fords, GM Holdens (Aussie Vauxhalls), Toyotas, Ferraris, GMCs (American truck things), Mercedes Benzes, Teslas, Jaguars, Land Rovers, Volkswagens, Audis, Hino Trucks and Skodas.

Recalls are, on the face of it are all about making cars safer by updating, repairing and modifying them. From the manufacturers point of view, it would be better if the vehicle could be replaced. Giving buyers a fairly spurious reason to do so, especially if the car isn't broken at all, would be their goal. Quite simply they would like some encouragement to buy the latest model. Every year there are different colours and extra widgets to make your motoring life so much better. It is called marketing and is promoted through advertising. Essentially this is telling fibs on a grand scale.

Back in 1960 Vance Packard called out this practice in his brilliant book 'The Wastemakers'. He rightly said that this was "the systematic attempt of business to make us wasteful, debt-ridden, permanently discontented individuals." He was well ahead of the green movement with his conclusion that disposable consumer goods was degrading the environmental, financial, and spiritual character of American society.

Cars, or automobiles as they are called in America actually gave him loads of material and arguably the perfect examples of planned obsolescence. He identified that in 1956 "the motorcars being led to the scrapyard chopping blocks were three years younger than the motorcars being scrapped in the late forties."

It is worth reading the book if you are into design as he explains how cars became larger, longer and lower, gained lights, lost lights, had dummy lights, grew wings then became stunted. It is all about style over anything actually useful, he called all these changes 'swings'. He said, "Every swing, whether seemingly sensible or seemingly insensible, helps create obsolescence. It is the swing that counts."

Interestingly old Vance did pick up on the fact that Brits were different, "A sales official of the British Motor Corporation Ltd was heard to complain: 'all you've got to do is drive out on a Sunday and have a look. You'll find people out washing and polishing cars that are twenty years old. Why, the blighters think more of their old cars than they do of their wives!' That was from a 23rd June 1960 edition of the Wall Street Journal, so it must be true, that and the fact he used the word blighters. This is proof that way back then Brits were enthusiastic practitioners of Bangernomics.

The forces of industry and control have always worked against the freedom of drivers to continue as they were. Just imagine if car manufacturers could get together with the government and persuade them to help subsidise a scheme whereby an old car is replaced with a brand new one?

Say hello to the 2009 scrappage scheme, which arrived as a result of the car industry begging the government back in November 2008. By then car registrations had dropped by almost 37 per cent compared to the previous November and some specific manufacturers, like Jaguar were down by more than 60 per cent.

A scrappage scheme was introduced on 18 May and ran until 28th February 2010 offering a £2,000 discount on a new car or light commercial vehicle (van up to 3.5 tonnes). It was underwritten by £300 million in taxpayer funds with that amount matched by the manufacturers, so in all £600m was up for grabs.

There were terms and conditions of course. The vehicle must have been first registered in the UK before 31 August 1999, to the person making the claim and owning it for the previous 12 months. It needed to be MOT'd, taxed and insured. However, if the MOT had expired, it must have done so no longer than 14 days before the order was placed for the newbie.

As well as drumming up business for the car trade suffering after the 2008 world economic downturn, it was justified on the grounds of cleaning up the environment with more environmentally friendly (their words) and safer cars. Officially they used data from the Society of Motor Manufacturers and Traders (SMMT), 85% of a car's life cycle CO2 comes from the use phase of a vehicle. Taking an average mileage of 12,000 miles a year, and exhaust pipe CO2 emissions of 165g/km, apparently it would take one year to pay back the emissions created through production, which account for 10% of lifetime CO2.

Also pre-1999 vehicles had Euro 2 engines if they were lucky compared to Euro 4 at the time. The stats say that those engines delivered more than a 50% improvement for harmful emissions and should need to prove this has happened in order to claim back the scrappage fee.

The official line also was that when a car is scrapped, 85% of it can be recycled or re-used. Once reusable parts have been removed the vehicle is then de-polluted by removing its harmful liquids, tyres and deploying the airbags. The remaining materials are then shredded to recover valuable parts like steel, aluminum and recyclable plastics, some of which go back into vehicle production.

Despite all this propaganda the point was that in many cases perfectly good MOT'd vehicles with many years life left in them were being scrapped. Never mind that some historically important cars were sacrificed under the scheme. There is a depressing list which includes some incredible classic cars such as a Lancia Delta Integrale. Clearly the owners didn't care enough to save them and the car dealer had their eyes fixed on the tax payer subsidised £2000 payback which makes their lives easier. Not only that, there may have been a manufacturer backed new car registration incentive to reward a dealer for hitting sales targets.

Well, the scrappage scheme did have the desired effect

and 396,000 new cars were sold in that period and there were eight consecutive months of growth with car production was up 62.7 per cent by the end of the scheme. It was calculated that 20% of sales were attributable directly to the scheme. In reality, buyers were simply getting a better than average part exchange deal. Well, that might have been the case in some scenarios, but is also quite possible that many actually lost out. If people were stupid enough to exchange classics which were obviously worth more than two grand, even in a shocking state, who knows how many bad deals there were.

The justification for all this was that the air would now smell of freshly mown grass and be filled with fairy stardust. Plus because these cars are newer and safer, less people would be killed or injured although that would be a lot harder to prove. Unfortunately the environmental argument does not stack up. With these exchange schemes, money and a new motor for a ropey old car, just moves the pollution somewhere else. In particular that would be areas around factories where the new cars and their components are made.

The timing of this scrappage scheme also meant that many of the brand new cars were actually the less than clean diesels, which everyone is now complaining about now. See the diesel chapter.

The upside for the government was that it in theory didn't cost them anything as the amount being paid out was more than made up for by the increased amount of

VAT paid on new cars.

All this showroom activity meant that members of the public who had previously stuck to buying used cars could now justify something brand new. Effectively the Government helped to give South Korea a massive motoring foothold in Britain. What had been justified as boosting the parts industry in the UK just seems to have been an exercise in recycling money. Hyundai sales rose 102 per cent from February 2008 to February 2009 and its subsidiary Kia, saw the scheme account for 70 per cent of its registrations. Other winners were Ford, Fiat, Toyota and Skoda and out of those only Toyota made complete cars in the UK.

Not every part of the car industry did well. For used car dealers the scheme actually reduced sales of their newer and often more profitable stock, in particular that was cars which were one to three years old and overall used car sales in 2009 fell to 6.8m down 5.7% on 2008. Meanwhile, new car dealers seemed happy and associated businesses like the companies which ask you half an hour of questions in order to let you have insurance cover, was also a lot busier.

A successful scheme for all those concerned it has not be tried again although there have been scrappage schemes in Europe, often with more generous terms. But just when you thought the tax payer backed scrappage scheme had gone away and there would only be those marketing led manufacturer ones...it came back to guess where?

London became the focus for a £25 million scrappage scheme in October 2019 was launched to take older cars and motorbikes off the streets of London. It was open to low-income and disabled car owners who live in Greater London. It was based on a scheme which was started in 2018, with £23m put into a fund to replace diesel vans.

Now, eligible Londoners could scrap vehicles in exchange for a cash grant of £1000 for a motorbike or moped and £2000 for a car.

In order to receive the grants, the scrapped vehicles must not meet ULEZ emissions standards. Owners must live within either the City of London or one of the London boroughs and also receive benefits. Vehicles must have been registered to the applicant for at least 12 months, be insured and have a current MOT and road tax. That grant money can be used only to buy a vehicle that meets ULEZ standards.

It is being sold by the London mayor as part of the fight against pollution, but quite how someone on benefits can afford to buy a brand new car is not really addressed. That's because there is no requirement on the recipient to purchase another bike or car. The vehicle to be scrapped must be taken to an approved scrap facility, by the owner, and they must provide a certificate of destruction in order to receive the grant. Those that qualify will, for example, receive £2000 of Zipcar credit (car sharing scheme), and free membership of the cycle hire scheme where all journeys up to 30 minutes are free.

On the face of it this is all about changing attitudes and clearing the decks of dirty old cars. It could also be another way of getting those who most need access to private transport to fall back on public transport. Targeting those with motorcycles seems especially misguided and who knows it may actually prevent them from finding work, or visiting their mum.

Perhaps this is the future of scrappage, just to get you off the bike and out of the car. It is in short, a bribe.

*The 'All-New' Top Gear Team in 1886*

Sucking up all the bad stuff...

## 11. The Catalytic Con - *Smoke and mirrors...*

It is always difficult to get away from the fuel, which we put in the tank. In particular, petrol and diesel has proved to be problematic.

What comes out of the exhaust pipe isn't very good and that's why if you want to end it all, then running an engine in an enclosed space is tragically all you need. That's right, unleaded petrol isn't exactly a clean substance. It does seem that we have not been told the truth about what's actually in petrol and misled about the best way of cleaning it up.

Let's hold our breath and just recap at what comes out of the back end of an internal combustion engine. Carbon dioxide gets all the attention but there are others. Carbon Monoxide, which is colourless, odorless, tasteless, yet highly toxic. Nitrogen dioxide is toxic by inhalation and can cause adverse health effects at low levels over a long period. Sulphur dioxide can cause pulmonary and respiratory distress and acidification of waterways.

Particulate matter, essentially soot, affects breathing and respiratory systems and causes cancer. Benzene is a carcinogen. Formaldehyde is another carcinogen. Not good. Back to the story.

As already mentioned, internal combustion engines originally ran on motor spirit. This is not just a quaint old term from the turn of the 20th century, this was properly pure stuff. As a result, what came out of the exhaust was water vapour, carbon dioxide and some trace carbon elements and particles. If that were all that came out of an exhaust these days, everyone and that includes the environmental lobby, would be organising street parties to celebrate.

The trouble with motor spirit was it cost a lot to make because this nectar like fuel was so refined. Oil companies wanted to make serious money by selling lots of it and not putting off buyers with a prohibitive price. Yet as compression ratios were raised to improve performance the lower quality of fuel could not cope and the engine would run very poorly, the technical term is pinging. The simple and cheap solution was adding some lead (actually lead tetraethyl) silenced the ping, hence leaded petrol.

Unfortunately over the years high-grade, low-sulphur, low-nitrate oil, which made the best petrol, was becoming scarce. So companies switched to cheaper high-sulphur, high-nitrate oil. Unfortunately this resulted in higher levels of sulphur dioxide and nitrogen oxide in vehicle emissions. Smog was created and made people cough and become very annoyed, especially in high-density automotive areas like Los Angles.

The time was right to start bolting the catalytic convertor to the exhaust end of automobiles, (American for cars). The oil companies were expected to do something, but obviously taking all the contaminants out of their refined products would cost money and quite possibly mean that refineries would have to be refitted. Better to pass that responsibility on to say, the car companies, who were getting most of the stick for the smog anyway. Rather than sticking a giant catalytic convertor into a refinery, fit millions to cars. Boffins, or very specifically a French engineer living in L.A. named Eugene Houdry, had already invented Cats.

There were experiments back in the '50s to filter out CO2 from power stations so that the gas could be repurposed to stimulate plant growth, and yes there is a Demotorized chapter on all that. However, the other pollutants remained a problem. What Houdry found was that passing gases through a filter made of platinum caused a catalytic conversion of the oxides to other products which could then be prevented from escaping. It was a costly solution, which did not operate for very long and anyway some of the by-products could be worse than the harmful oxides. There again it only had to work for a while with cars to convince the majority that this was the answer. Smoke and mirrors?

Apparently smog could be stopped in its tracks by these clever cats. Except that the lead in petrol would kill one stone dead. Alternatives to lead, which you may remember stopped the engine going 'ping', led to even more terrible things to come out of the exhaust pipe. Luckily a catalytic convertor could deal with them. This is when the oil companies needed to get really clever and call in their Public Relations experts.

What they had to do was scare everyone into believing

that the lead in petrol was life alteringly dangerous and should be replaced with a brand new unleaded version. The inevitable consequence would be legislation, that would lead (a pun I suppose) to banning leaded petrol, which would clear the way for the adoption of catalytic converters as standard.

As a result the significant smog problem was fobbed off onto the car buyer who would ultimately have to pay the higher price of fitting a catalytic convertor to their car so they would be able to drive to work, somewhere in Los Angeles. What happens first in America inevitably drifts eastwards to us, more on that later.

**Catalytic Converter sort of explained** The point of a cat is to burn unburned hydrocarbons (fuel molecules) left over from the combustion cycle and worked well enough with old carburetor engines in the mid '70s when the fuel mix ratio was set to the best compromise for most conditions.

The problem with early converters is that they took a long time to get hot, and they had a problem staying hot enough to burn the fuel all the time. This led to the invention of something unique to America called the smog pump. This blew air into the cat like a blast furnace to keep it hot, and the carburetors were set even richer to make sure they wasted enough fuel to keep the cat hot at around 400 degrees C. That cost engine power and used even more fuel, which seems daft, but using more fuel to burn unburnt petrol would not have troubled an oil company. Or bother the people who made the cats and that was mostly the Merchant Bank to precious metals dealer, Johnson-Matthey Limited.

A report in the July 1986 edition of the New Scientist revealed that British ministers called Catalysts a 'cul-de-sac' technology. They wanted to cut emissions by making engines more efficient with a lean burn technology. These changed the mix of air and petrol at the point of combustion in the engine. There is a complicated, technical explanation, but a combination of a modified cylinder head and digital ignition technology, burns more of the fuel, more efficiently.

More interesting is that Europe's catalyst manufacturers formed a group called Automobile Emissions Control by Catalyst (AECC). The British representative at the time, Robert Searles said that the claims by manufacturers and the Department of Transport about increased fuel consumption, reduced performance and higher prices were unfounded. He had a point that a 25% increase in fuel use was wrong because it was much nearer to 4% and top speed was trimmed by a marginal 3 kilometres per hour. As for the price of a brand new cat equipped vehicle, the average increase was £214, although some companies like Volvo and Fiat did not charge any more for a 'cat' model.

It was believed that a lean burn engine would have met the standard EEC test drive. However in the real world there was evidence that they would do worse as the average required speed was just 18 kilometres per hour. That test never happened, so we don't know. What we do know is that there was now a very effective pressure group with an almost bottomless war chest to promote the catalytic agenda.

One argument is that since the introduction of very sophisticated, computer controlled fuel injection systems, which precisely deliver the right amount of fuel to the engine, would mean that no modern engine needs a cat. All the catalyser actually does (certainly in the early days) is use up to 20% of the fuel to keep it hot. Even so, the car releases more hydrogen sulfide into the atmosphere and there is the ongoing issue of lead. No cat means lead could make a comeback, but would that be wise?

Of course lead is bad for you, so bad that Nigel Lawson plonked a surcharge on it in the '80s. Conclusive proof that lead is really bad can easily be found in papers like the snappily titled publications from the American National Academy of Sciences/National Research Council: Measuring lead exposure in infants, children, and other sensitive populations. Washington: National Academy Press; 1993. That concluded that lead is toxic to multiple organ systems. The nervous system of the foetus and infant is especially susceptible to lead, which can cross the placenta and penetrate the blood-brain barrier. Lead interferes with neuronal migration, cell proliferation and synapse formation during critical periods of early vulnerability. So it is bad for babies and potential babies when they are still in the womb.

However, the evidence is that getting it from an exhaust pipe and into the human body is not clear. It is possible to measure lead as grams per cubic metre by the side of the road with a monitor, but that will fall quickly as lead dust is dense, small and heavy, so not immediately breathable. Monitoring air quality around the world seems to bear this out. Germany were enthusiastic adopters of the unleaded agenda in the 1980s and Frankfurt's Federal government cut the lead in their petrol from 0.4 to 0.15 grams per litre. Yet they measured nearly a thousand people over a five-year period and there was no change in the amount of lead in their system.

Research from the University of London found that the lead coming from the exhaust had been cooked at 2,000-3,000 degrees centigrade in the engine. The particles are heavy, don't become airbourne and can only be seen with a microscope and don't get absorbed into the body. Lean burn engines would have changed the mix of air and petrol at the point of combustion. If there is any lead in the body it has not come through the air. Yet the narrative has always been that lead makes children thick. That is certainly true, but what is, or was the source of the lead?

Let's turn to Australian scientist Dr David Warren AO, BSc (Sydney), PhD (London), DIC, DipEd (Melbourne) whose research showed that the lead in blood comes not from breathing airborne lead but from eating and drinking it. That happens when humans use soldered food containers, lead-based paints and lead pipes. Additionally research showed that the blood lead levels were higher in countries where people drank bore water, such as the New Guinea highlanders and people on remote islands. There were virtually no cars in the locality, but their blood samples contained more lead compared to those living in the heart of a big City, in that case Melbourne, Australian.

Lead is bad for people, that is certain, but it does not seem to be conclusively proven that lead in petrol is. Indeed, the aromatics added to petrol to make up for the loss of lead constitute much more of a threat in themselves. These include benzene, toluene, xylene, mesitlylene and dimethylbenzene which could be as risky or even more dangerous than lead. Benzene is rated as one of the most toxic and carcinogenic chemicals in atmospheric pollutants. Unleaded petrol is really, really bad.

With lead replaced, the pollution actually became worse, as those aromatic compounds are very photochemically active and there was more of the known carcinogen benzene around.

In 1994, a group of MPs called on the government to ban sales of super unleaded petrol. They said that evidence "strongly suggests that the potential health hazards resulting from the excessive aromatics used...outweigh any possible benefits from the reduced lead." They also wanted the composition of petrol to be publicised so that people could compare the different types of fuels. Dr Simon Wolff, from University College London School of Medicine, convinced the House of Commons Transport Committee that unleaded petrol, especially super unleaded, replaced lead with yet more dangerous levels of cancer-causing aromatics. That did at least result in a higher tax on super unleaded.

Originally Wolf was concerned about pollution from Britain's nuclear power stations. Whilst doing this he noticed a tenfold difference in childhood leukemia rates between certain populations. He concluded that newer middle-class suburbs and towns with high levels of car ownership were the most at risk from the disease.

According to Dr. Wolff, who died in 1995, the country should be aiming to lower the lifetime risk of leukemia to one in a million, so benzene levels would need to be cut by fifty- or one-hundredfold. The reason why children are more vulnerable to lower levels of benzene is that compared to adults, they breathe faster, have a much quicker metabolism, and their bone marrow is more sensitive.

He was backed up by research from the Environmental Protection Agency in America EPA, which claimed that half of all cancer cases might be related to air pollution. A lifelong exposure to a microgram of benzene per cubic metre, roughly 1 part per billion, they estimated that 2.8 people per million would be diagnosed with leukaemia. That means that it is unleaded petrol, which could still be killing us.

That's an uplifting thought, so here's an interesting diversion from 1993. Plain old Mr. James Dyson as he then was popped up on BBC TV with a device which was certainly a lot simpler, cheaper and more effective than some old cat. He called it a cyclonic vehicle exhaust, which was the size of a tea flask and could be inserted into an existing system and trap 95% of the most harmful emissions. He is shown tapping out the black grit onto a white piece of paper. Talking to The Independent newspaper at the time about the 20 million vehicles on the road, "Every one produces six grammes of dirt for every hour that it has been used. That's a huge amount, which goes on to cause cancer and acid rain. I can stop it."

That was a very bold claim, but it was a simple vacuum based solution by a talented engineer and not a complicated precious metal concoction reliant on some global multinational companies to do some profitable mining and marketing.

If conventional fuels are so bad then there are some more credible renewable options, such as biodiesel and ethanol, which have been blended with petrol and diesel in the UK for the past 10 years. The point of them is that they lead to a reduction in overall carbon dioxide emissions and helps the country meet climate change targets. Apparently, blending renewable fuels in this way contributed to a CO2 emissions reduction equivalent to taking over 1 million cars off the road. The labelling isn't terribly clear, but the petrol badged E5 (ethanol up to 5%), whilst diesel has a B7 (7%) tag.

Then there is E10, a biofuel made up of 90% regular unleaded and 10% ethanol – hence the E10 name. Unlike regular unleaded petrol, ethanol absorbs carbon dioxide, partially offsetting greenhouse gas emissions they say. This fuel is corrosive to rubber parts on older cars, so should only be used with expert advice. However, the Society of Motor Manufacturers and Traders (SMMT) estimates that 92.2% of the petrol-engined vehicles in the UK are compatible with E10, but the remainder are not. As of 2011, all new cars sold in the UK must be E10 compatible.

Drivers of cars registered prior to 2002 were advised not to use E10 in their vehicle, as problems have been reported. Research carried out by What Car? revealed that E10 is potentially less efficient than the current E5 blend of fuel, with the problem being worse in smaller-engined cars. So drivers of shopping cars would end up filling their cars more often, which is not the point of owing a small car with a tiny engine.

Cats are legally here to stay and are unlikely to go away, unless you buy one of those brand new EV things. With those, the bad stuff may come out of a power station. Maybe then we should go back in time and plonk a great big Cat or two on top of those.

Dieselmonkey

**12. Dieselgeddon** - *How the world fell out of love with Derv.*

If you burnt diesel in your vehicle back in the old days, chances are you really did work hard for a living driving bus, lorry, train or a digger. Then diesel became socially acceptable as technology civilised the fuel of the masses and politicians gave drivers encouragement to put oil in their tank. Apparently it was all a great big misunderstanding.

The official origins of diesel date back to 1892 when Rudolf Diesel, not a coincidence, worked his socks off by patenting a combustion engine that could convert most of the heat to useful work. It was far better than the engines around at the time, and it used up a by-product that otherwise would be thrown away.

Except that there is a British strand to this story as Herbert Akroyd Stuart, working with Richard Hornsby and Sons, patented the so-called "heavy-oil engine" in 1890. It was a fuel-injected unit that ran on paraffin oil, also known as the 'hot bulb' engine, so needed some pre-heating, making it not as straightforward as Diesel's.

Anyway, things didn't go well for Diesel and if you enjoy a good old conspiracy theory he expired in very mysterious circumstances in 1913. Diesel had to cross the English Channel for a meeting in London about a plan to produce and install his engines in Royal Navy submarines. Except he disappeared overboard and it was ruled as suicide. The thing is though there were some industrialists who were not keen on this rival technology. Diesel engines were not just efficient they would run on pretty much anything that could be loosely defined as a biofuel, such as peanut oil. Consequently oil barons were not best pleased, and with the First World War just a year away, perhaps the German were concerned that Diesel was fraternising with a future enemy.

Diesel like so many clever inventors, never made any money from utilising a petrochemical bi-product and never had a Deutse Mark to his name. Except that he did give his Mrs. a suitcase full of 20,000 Marks just before he got on the boat. There was clearly something complicated going on, but that's nothing compared to what would happen a century later.

The diesel engine was used mainly as an industrial power unit, doing most of the heavy lifting in life, be it a lorry, van, taxi or taking over from the steam locomotive. They were rarely used in private vehicles unless the price and tax regime was favourable, usually in Europe. The simple fact was that a diesel was not very sophisticated, relatively slow, noisy and smelly.

However, diesel started to gain some serious traction for cars when the 1973 fuel crisis was in full swing. It seemed like the perfect answer when all motorists needed was to get more miles to their gallon of fluid and by then they didn't care what it was. Also, diesel engines were tougher and could do bigger mileages.

Some of the American V8 diesel reworkings were not great and could have been more reliable. America's heart though wasn't really in it and when the fuel crisis was over, everyone went back to using petrol powered internal combustion engines.

Meanwhile In Europe the race was on to make diesels socially acceptable and much classier. The technology improved as turbocharging was used to give the engines better performance along with direct-injection in the 1990s. Common rail systems where the fuel is stored at variable pressure in a cylinder or 'rail' connected to the engine's fuel injectors via individual pipes. This provided a level of flexibility, leading to better emission control, power and fuel consumption. This also made the engines much quieter than ever before.

The problem with diesel all started with the Kyoto Protocol in 1997. That's when governments particularly European ones signed up effectively for their own version of the diesel new deal. It was an easy way to meet their CO2-reduction targets.

Diesels emit all sorts of nasty things like nitrogen oxides and particulates. However, they overlooked all that. What mattered to the idiot politicians was that CO2 emissions were much lower than petrol engines and that had been identified as a greenhouse gas, which they believed, contributes to global warming. Well it was called global warming at the time.

**Diesel Fallout UK Part 1** That's why governments also

spent billions subsidizing diesel and fiddling with the tax rates. Chancellor Gordon Brown brought in a sliding scale for car tax in the 2000 budget, to make it cheaper for cars with lower emissions of carbon and lower tax rates generally for diesel cars.

Reading the advice given to the ministers at the time from the Treasury's civil servants they said, "Relative to petrol, diesel has lower emissions of CO2 but higher emissions of the particulates and pollutants which damage local air quality. A diesel supplement is necessary so that we do not create incentives for people to choose diesel vehicles over similar petrol models in order to attract a lower VED rate."

Despite that, the politicians didn't want to look bad or penalise diesel users, so settled for a token £10 supplement. It made sense though as the resulting financial incentive for diesel cars helped to boost sales once the new tax laws arrived in 2001. At the time there were just 3 million diesel cars registered in the UK. By 2016 that had risen to 12 million as a direct result of those Government incentives. Suddenly, diesels were half the car market.

Diesel emissions of nitrogen oxides and particulates have been linked to respiratory difficulties, heart attacks and lung cancer. Indeed, The World Heath Organization published a report in 2012 saying just this.

The subsequent Conservative Government when Michael Gove then Environment Secretary used it as a party political stick to beat them with and said, "The dash for diesel was pursued under a Labour government, and these documents show they knew the damage this would do to our environment. This is yet another example of a Conservative government having to clean up Labour's mess.

"We are taking action and have put in place a £3bn plan to improve air quality and reduce harmful emissions as well as ending the sale of new diesel and petrol cars and vans by 2040."

Clearly one bad idea has been followed by another misguided one which brings us to those naughty Germans at Volkswagen.

Saying something is a gate, when it clearly isn't, is not very original. I always prefer a good Dieselgeddon, but Dieselgate it has become.

There are vehicle tests to establish emission outputs and overall fuel economy. These tests vary from country to country, but essentially they all work in pretty much the same way. A simulated set of procedures involving acceleration and braking regimes, which are designed to mimic driving in town and driving on a motorway. It is all made up and rather artificial especially as it is conducted in a controlled environment. Not surprisingly, all manufacturers have routinely cheated to make their figures more customer friendly.

To this end they overinflate tyres so they had less rolling

resistance making the vehicle would go further. Not only that, they would tape up panel gaps around the bodywork to make them more aerodynamic and so use less fuel. It didn't end there, removing any superfluous bits of trim and spare wheels, when they existed, to make the model lighter would also boost economy and performance.

When it comes to modern day emissions, the only way to fake the results is to interfere with the actual engine electronics because these tests are conducted in a laboratory. Cheating software was therefore programmed to detect steering, throttle, and other inputs used in the test so that it could switch seamlessly between the two distinct operating modes. What would happen is that the software would significantly change the fuel pressure, injection timing, exhaust-gas recirculation, and, in models with AdBlue, the amount of urea fluid sprayed into the exhaust. While this mode delivers higher mileage and power, it also permits heavier nitrogen-oxide emissions (NOx).

This brings us to September 2015 when Volkswagen was accused of installing this sort of software to cheat the test in America. The history of this stretches back almost a decade, it was all part of a dastardly plan apparently.

Back in 2006 Volkswagen decided to follow the money, to the biggest new car market in the world, America.

Their core engine was diesel, a technology they knew a

lot about and had invested so much into. The alternative was going down the hybrid, petrol/electric one favoured by Toyota and Honda. Taking the diesel route meant being a bit sneaky, as a diesel was never going to reach the rising emissions standards.

Volkswagen's 2008 marketing plan cheekily positioned their models as 'clean diesel'. They even went to the bother of building a factory in Tennessee to pump out their 'clean' Passats.

Things started to unravel in 2013 when West Virginia University got a grant from the International Council on Clean Transportation. They used that grant to compare real world emissions of diesel cars rather than rely on the existing laboratory tests. What swung it for them was finding a couple of the VWs, which under their evaluation criteria emitted more nitrogen oxides than, would have been permitted under the tests.

As a result of those findings the Californian Air Resources Board opened an investigation in 2014. Volkswagen was not especially helpful. Eventually they did issue a recall in 2015 and said they would issue a software to fix what they regarded as a technical anomaly. Meanwhile, back at the corporate office, documents were being shredded, presumably by VW interns.

After unrelenting pressure VW had to admit that they did fit insanely clever defeat-device software. The sort of software, which knows when it is being put to the test, triggers the pollution controls to ramp up and comply with the emission regulations for the required period. Some half a million vehicles were fitted with the dodgy devices in America, but worldwide that was a much more considerable 10.5 million.

That emission admission led to VW boss Martin Winterkorn resigning, but not before blaming staff that had gone rogue. That old one. Eventually a few suits got prosecuted and two were actually handed prison sentences in the States.

In 2016, billions of dollars started to flow from the Volkswagen bank account in compensation to owners and to do environmentally friendly things, like install charging points for manufacturers who actually made EVs. So $14.7 billion was paid out along with a further $4.3 billion because they broke the Clean Air Act. It didn't end there because VW own a lot of brands so a further $1.3 billion was paid out to owners of Audis and Porsche, plus owners of VWs with the biggest 3 litre engines. And there was even more money, because Robert Bosch who actually supplied the cheating software, paid out $327.5 million to owners. Incidentally, we are still in America at this the point, the land were there are 1.35 million lawyers according to Statista.com.

As well as compensation there was a buy back scheme which parked VWs pointlessly in compounds. There was a software fix offered free to owners that wanted to keep their VWs who also got compensation for reduced resale value. So hundreds of thousands of perfectly useable (and arguably cleaner than all of the locally made petrol V8s) VW cars in USA were stockpiled for scrapping. This was an even bigger environmental disaster than failing the test, or if you were feeling generous, bending the rules.

This is all very fascinating but what is wrong with this billions of dollars compensation scenario? As mentioned, it is all North American. That's because Canada got a handy $1.6 billion package.

Even in Australia at the end of 2019 Volkswagen Group and Audi settled two major Australian class actions. Around 100,000 VW, Audi and Skoda cars were affected and after denying it had a case to answer the companies agreed to a pay-out of up to $127.1 billion, about $1,400 per eligible car.

Meanwhile in Europe, nothing. Not a penny offered. Just a billon Euros in settlement with German prosecutors. Yep, Germans are very good a being sorry, without actually being at all sorry. Consumers don't see a problem and have just kept on buying their cars. In 2018, the Volkswagen Group set an all-time record with 10.83 million vehicles delivered throughout the world, that's 0.9 percent more than the previous year. Not only that, their operating profit for the year was 13.92 billion euros, 0.7% better than the previous year.

Volkswagen is now off to reinvent themselves as a

battery car builder and hopes that we will forget about all this and incredibly that strategy seems to have worked.

It isn't just VW who wanted to cheat the system and mislead customers. Lots of other cars have been found to have worse emissions in real world conditions rather than in the lab. Specifically. BMW and Mercedes-Benz had their offices raided by investigators searching for evidence. The European Commission has been looking much more closely at whether VW, Audi, Daimler, BMW, and Porsche actually got together and colluded right back in the early '90s when diesel engines started to get really good. These companies have been accused of forming an illegal cartel, "If proven, this collusion may have denied consumers the opportunity to buy less polluting cars, despite the technology being available to the manufacturers," said Margrethe Vestager, the European Commissioner for Competition.

In 2019 German prosecutors handed down a fine to BMW as well. BMW will pay $9.6 million (8.5 million euros) for installing "defective" engine software in their diesel cars. Those prosecutors found that the company mistakenly installed incorrect software in 7,965 cars, and fined the company for lax oversight and insufficient quality control, but said they found no evidence of fraud, as they did at Volkswagen and Audi. Maybe BMW were a lot cleverer, had better lawyers or had seen what had unfolded at VW before putting their defence together. No wrongdoing, just lax quality control, not something BMW have be known for, even if it does relate to just eight thousand cars.

Oh, and then there were the monkeys. Did we mention the monkeys? There were experiments, commissioned by the European Research Group of Environment and Health in the Transport Sector, a body funded by Volkswagen, Daimler and BMW. Thanks to some digging by the New York Times, the actual tests were conducted an awfully long way from Europe. It all happened in May 2015 and was carried out by the New Mexico-based Lovelace Respiratory Research Institute.

They used ten Java monkeys, contained in small airtight chambers for four hours. Left to watch cartoons as they breathed in diesel fumes from a VW Beetle and later on an old model Ford F-250 pick up. There really isn't anything very funny to say about such a grim situation. VW and their mates at BMW and Mercedes wanted to establish that nitrogen oxide wasn't the problem.

The evil doctor types, presumably not German then examined the blood, lungs and bronchial tubes of the monkeys for signs of damage, but it didn't end there.

According to reports in The Süddeutsche Zeitung the experiments were also carried out on 25 young people. Not sure if they were restrained, put in airtight boxes or forced to watch Tom and Jerry, but this happened in Europe at the University Clinic Aachen. So the doctors were quite possibly German.

Not sure if all this was worth it, as the findings haven't been published. Someone somewhere probably knows. No monkeys seem to have be involved with all the other allegedly cheating companies. Fiat Chrysler settled a case in the U.S. over its Ram pickup and Jeep Grand Cherokee EcoDiesel models, which also used software from Bosch.

Meanwhile, the French authorities have looked into VW, Renault, Peugeot and Fiat. In particular the investigation by experts at the University of Burgundy found that two systems on the Renault Captur and Clio Mark IV models were designed to filter emissions and didn't operate at many speeds, or set outside temperatures. These findings could put Renault among a host of European carmakers to break rules on controlling emissions for cars running on diesel. Presumably they will get fined or something, but the actual consumers who have been misled won't get very much at all in the way of compensation. Apologies cost nothing. Perhaps it will just be forgotten about.

Here then is conclusive proof that whatever car manufacturers say about emissions, the planet or being straight with their customers, they are less than truthful. Manufacturers do not care about any of this. They are prepared to cheat the system, lie about their environmental ambitions and credentials, just to sell cars. Never forget that.

Then there is WLTP (Worldwide harmonized light-duty vehicle test procedure) used from September 2017.

Miles Per Gallon figures (how far will a car travel on a gallon) have always been downright lies in most cases. Under the old New European Driving Cycle (NEDC), cars on a rolling road accelerated, maintained a steady speed and braked over a 20-minute cycle.

A British Company, Emissions Analytics, independently tested hundreds of vehicles, covering a wide range of makes, models and engine types and created an Equa Index. They found that the average car's fuel economy is 25 per cent lower than official numbers, whilst some were 40 per cent below the original laboratory figure.

It should be pointed out that the new WLTP is still laboratory based. It includes different average speeds, ambient European temperatures, so cold for Britain then, stop-start, like you are driving for a courier company and testing different models in a vehicle range, from poverty to rich specs. These are early days and the new system won't be used to calculate CO2 and therefore road tax until April 2020. Manufacturers though will be able to quote both sets of figures, which won't cause any confusion at all.

**Diesel fallout Part 2** Back to Blighty and it really is no surprise that in 2017, diesel sales in the UK plunged by 17%. With diesel bans coming to European cities, motorists are anxious about being stuck with a car that will become unsellable or unusable.

When it comes to bans everything always leads back London. Here was something created especially for those lucky diesel drivers and the ULEZ D-Charge. This was an additional parking tax for diesel drivers in London. Drivers of certain diesel vehicles paid up to £50 a day (around £2 to £2.45 an hour) if they were planning on parking in certain areas of central London.

Yes, the future does not seem to be very bright for diesel cars. This is repeated elsewhere, but Bristol has made a name for themselves with a proposal to ban private diesel cars. Except that Bristol Council admits that EU6 diesel taxis are compliant and won't be subject to its proposed Clean Air Zone restrictions. That is a bit perplexing if you are a resident of the area with a relatively new privately owned EU6 diesel car that will still be banned. Such rampant local authority hypocrisy has not gone unnoticed when it comes to their double standards on diesel.

What has also been killing diesels apart from the manufacturers, the government and local authority regulation is the technology itself. Never mind that when you fill up you can get the smelly, slippery stuff on your hands especially if you stupidly don't wear gloves.

In order to comply with emissions requirements, actually using diesels has become a real pain. First of all there was the Diesel Particulate Filter (DPF). It was specifically designed to reduce the amount of soot and ash passing through the system in order to clean up the exhaust gases. A diesel particulate filter will gradually become restricted and more clogged as the miles build up. That can lead to a loss of power and potentially the engine could overheat. Then there is the EGR Valve, which is designed to recirculate exhaust gases back into the engine before they pass through the DPF. If the EGR is not working effectively or blocked, the gases going through the DPF will be increased which will contribute to a blocked DPF. Alternatively if swirl flaps are fitted which perform a similar function to an EGR, when not working effectively that can also block the DPF.

The trouble is sorting out a blocked DPF is not cheap. Sometimes they can be cleaned, but replacement can sometimes run into thousands of pounds. Driving slowly around town has been found to be the best way to kill off a DPF. That used to be something a diesel engine was good at.

Then there was AdBlue. This is a fluid made up of a mix of urea (that is a really posh, or medical way of saying wee) and deionized water, which is squirted into the exhaust system, helping to reduce nitrous oxide emissions produced by diesel engines. This is called selective catalytic reduction, So as well as topping up with water and coolant, here is another fluid to worry about and pay for.

Meanwhile manufacturers are facing EU's stringent 2020 emissions regulations—or face huge fines. They argue that diesel technology is needed to bridge the gap until it's possible to fully migrate to electric vehicles. There are a lot of problems with that sentence. Currently the working world is powered by diesel. Lorries, buses ships and bulldozers to name just a few.

Yet the private diesel owner is the one who is being persecuted. Constantly.

What could save diesel? There is something called biodiesel which is renewable and clean burning. Biodiesel is typically created from vegetable oils like animal fats and unlike diesel is classified as one of the most eco-friendly fuels in existence. It is more than 75 percent clean fuel source. Engines that run on it are easy to start and to operate because of the high amount of cetane found in biodiesel. Oh, what's this? Palm Oil. That's not very good at all as a biodiesel. Think of the monkeys' and orangutans' habitat. Well that's more than Volkswagen, Mercedes and BMW ever did.

RIP Diesel. In cars anyway. Private buyers won't have a choice very soon. Thank you legislators and misguided environmentalists, but mainly the politicians who may well have access to a chauffeur driven official car, or can at least claim their diesel powered cab, on expenses.

Politicians were told that diesel cars produced less CO2 than petrol, so they must be better for the environment and that's why they encouraged them. Now we have the same thing occurring with EVs. Politicians believed that because an electric car does not even have an exhaust, it is emissions free. That's how stupid they are.

The car manufacturers are not stupid, simply dishonest.

*Robbie the Route Bus*

## 13. Rise of the Robot Cars - *Out to get you and possibly take you to the shops.*

Back at the dawn of motoring it was enough that these new fangled motors were 'self-propelled' rather than tugged along or about, by a horse. The thing was that a nag was seldom clever enough to take you where you wanted to go without some sort of human input. So it was with cars. The fairly complicated inputs required to make a motorcar start, stop and navigate to a particular destination required a human being. Not any more.

Why though would a so-called self-driving or autonomous vehicle be a good thing? Is it really what motorists actually want? Well, here are some of the arguments from the pressure groups and governments that want to put a robot in charge of your car. They are all strangely interconnected when it comes to benefits.

**Traffic Jams will be a thing of the past**. When cars are talking to each other and getting along famously, they will stay out of each other's way. Maybe it will tell you to leave the house at a different time. For a government that means less money spent on infrastructure and more control over who actually uses vehicles. Indeed, the road capacity is going to increase exponentially. Columbia University said switching to 100% self-driving cars occupying one lane mean that it could cope with 12,000 cars per lane per hour driving at a constant speed of 75 mph.

**Wow, so that means higher speed limits**? Not that you

will have any control over them, but the robot will be able to adjust the speed of travel in a nanosecond. You will be early for every appointment from now on. These higher speeds will only be allowed with electric cars, because of the environment and pollution and all that. Surely then there won't be **any need for traffic policemen** if everything is so under control. I suppose that's good.

**Even though there are higher speeds, there won't be any accidents**. That also means that the shape of future cars is going to be very different. As they won't crash then they could be made out of egg boxes, or something. That will mean that hybrid cars won't use very much fuel at all and those fully electrics will be using energy wonderfully efficiently.

As a result of all this inter car communication and cooperation **robots are so much better at driving** because they won't feel knackered after a hard day at the office, or be annoyed by a football result or some berk undertaking them. We are reassured by boffins that robots are so clever, pedestrians and cyclists won't ever be bothered by cars again.

**Finding a parking space won't be a problem either**. The car robots communicate with the robot running the parking slots and it will all work seamlessly. Driving into an area where there are no spaces just won't happen, say the robot's biggest fans, (other robots?)

The great thing is **motorists won't be wasting any of**

**their precious working time.** Rather like your boss can email you to write a report at the last minute when you are on the bus. Now the same thing can happen in your car. You won't be squandering time driving when you could be boosting your productivity. You will love this brave new driving world and so will your employer.

**Anyone holding out on this big idea of letting robots drive your car is just being selfis**. There are passengers who are more deserving of self-propelled and guided transportation than you. The elderly. The disabled. The children. People who can't or wont, drive could suddenly become far more independent. Imagine that, free from the school run, don't worry about old parents getting home after babysitting and blind Uncle Arthur can go wherever he wants for the first time in his life.

Indeed, this brave, new, self-driving car life is going to be wonderful. Before we get there, here is some background.

Actually, the whole concept of getting a car to drive itself and not rely on a fallible human is not new. The backstory of self-driving cars is fairly long and you can go down all sorts of self guided alleyways. In 1925 a Houdina Radio controlled car was navigated through New York traffic from a following car and became known as the Phantom Auto. However, full credit for imagining a fully formed and oddly quite logical self-driving future goes to Norman Bel Geddes. He took the German concept of the Autobahn and joined that to electronic speed and collision control systems found on railway networks. He brought these together in his Futurama ride for General Motors (GM) at the 1939 World's Fair when he imaged how things might be in 1960. What would happen is that drivers would arrive at the motorway, or freeway in America, then press a button and engage automatic mode. The cars would run in separated lanes. Leaving the motorway meant disengaging the system. Clever and still a possible future.

In the 1950s there were so called Autonomous Highway System tests, when GM and electrical giant RCA developed automated highway prototypes with radio control for speed and steering. Magnets in the car tracked a steel cable embedded in the road in Lincoln Nebraska and Princeton New Jersey. Here was a sensor-based system that could potentially have been rolled out nationwide.

Cruise control which is essentially standard on the majority of current cars and is even now, a fairly advanced driver assistance system was first fitted in 1958 to Chrysler Imperials. Self driving automobiles would seem to have been well on the way, but in reality robot cars would all be theory and engineers showing off for the next 50 years.

The whole idea of controlling the traffic has been very appealing to governments, which of course brings us to Britain. After not trusting drivers with unrestricted motorways the Transport and Road Research Laboratory conducted very advanced experiments from

1960 and into the early '70s in what was known as 'drive by wire' so that motorists no longer had to touch the accelerator or brakes. There was not a wire, like a tram, but there was a road and several miles of cable buried underneath it.

Although the British Laboratory used a number of vehicles the advanced Citroen DS19, was best suited to the system with its high-pressure hydraulic system modified to provide driverless operation of the steering, brakes and accelerator. It was equipped with an electro-mechanical self-driving system like the autopilot on an aircraft and connected to guidance cable/wire buried beneath the road by sensors. Indeed, some of that wire may still be below a section of the M4 between Slough a Reading where it became the world's first automatically controlled vehicle.

Controlling traffic is potentially an easy way of eliminating congestion and the experiments conducted by the Laboratory had established that it was possible for dozens of cars to travel inches apart as a 'train', without the discomfort, or curled British Rail sandwiches. Indeed, much is made these days of lorries and cars doing this quite soon, but potentially it could have been a non-rail reality 50 years ago. Back then Britain didn't have the money to invest in this, and was still struggling to build a decent road infrastructure. Maybe they should have forgotten about building the supersonic Concorde. Maybe it is a good thing they didn't, because true motoring freedom might have been surrendered a long while ago. It did prove that if you want to achieve some sort of autonomy, best to do it on a very controlled section of road like a motorway. Start to get fancy and clever by fiddling around busy town centres with people and bikes and it all starts to go wrong. Let's not get to far ahead and get back to the history of turning cars into robots.

Meanwhile, and still in Britain, antilock brakes which had also helped to bring aircraft to halt, by applying automatic cadence braking (rapid press/release style) was finally fitted a car, the rather expensive Jensen FF in 1967. It would help you stop sooner and most importantly drive around obstacles without the brakes locking up and skidding into oblivion, or a tree, pretty much the same thing. Unless you were a professional racing driver a robot is far better at braking than a human, but the driver still had to apply the more important inputs, the steering and swearing.

Returning to America the Space Race (Apollo moon shots and all that) where Stanford University built their so-called 'cart' to navigate around the surface of a distant planet. Fitted with cameras, it would evaluate the terrain and only move when it decided all was well. However, the Stanford Cart was not the sort of system you wanted to use to get to an important morning meeting in Leeds, never mind the moon, as it took quite a few hours to cross a small room dodging office chairs.

Over to Europe and fitting sophisticated cameras was the logical way forward. A professor at Bundeswehr

University, Ernst Dickmanns, took a Mercedes-Benz van and kitted it out with special 4D vision cameras and linked them to the steering wheel, accelerator, and brakes. What we have is a delivery van loaded with technology rather than parcels, or a driver. Apparently it was very good and came to the attention of the German government and the European Union. They said they wanted safer roads, or rather, roads they could control.

Eureka PROMETHEUS was a huge driverless vehicle project involving universities from all over Europe coming together with a massive amount of funding. A staggering €749m was sent into academia and that at least resulted in several cool cars. A Dickmanns converted Mercedes S-Class took a 1200 mile drive around Europe with a driver, but one who didn't touch the controls. Then in 1996 the Universities of Parma and Pavia drove their clever Lancia Thema 1200 miles around Italy, 94% of the time in autonomous mode. That technology never went away and the research and technology is still being evaluated today.

However, modified luxury Euro saloons anonymously criss crossing the Alps with academics, their equipment and the boot stuffed with hundreds of millions of subsidised Euros wasn't very sexy or interesting. By contrast in 2004, the U.S. Defence (I know they spell it incorrectly as Defense) Advanced Research Projects Administration (DARPA) challenged teams working on autonomous vehicles to compete for a $1 million prize.

The idea was that this would help the military's aim to make a third of its vehicles drive themselves by 2015 rather than have squaddies behind the wheel.

This best thing about this competition is that it made for some great film, which could be shared on You Tube when it started a year later. That's because in 2004, all the Robocars were rubbish and did not complete the real world desert course. Mind you, the boffins got their act together in 2005, no doubt inspired by the prize money doubling. First place was taken by 'Stanley', who was designed by Stanford University and Volkswagen. It might have looked like a Toureag, but it was related to that old Stanford Cart, which obviously didn't encounter too much randomly placed office, chairs in the dessert. Stan managed to negotiate 132 miles all on its lonesome.

We haven't seen any scary footage of terminator tanks engaged on manoeuvres on anyone's lawn, but something rather more worrying happened a bit earlier in the military timeline, which should possibly concern us all.

Letting academics and boffins at universities loose with the concept is one thing, as they really wouldn't know what to do with the end product, even if they managed to actually make one. Private companies though, need to monetize what they do. That's why BMW, Mercedes and even GM got back on board trying to crack the self-drive nut. The thing is, this sort of 4D Chess tech is prefect for the brilliant, but sometimes dysfunctional minds that inhabit Silicon Valley.

Presumably when the top Googlers, Larry Page and Sergey Brin searched for 'self-driving car'. they didn't like what they read. Maybe it was a picture of Elon Musk. We'll come to him in a bit, but let's get Google out of the way first.

The Google Self Driving Car project kicked off in 2009, because the company was always looking to do something else with all that money they were piling up, plus it was about the data. Once again plucky little Stanley edged back into the picture with the arrival of Sebastian Thrun, the Stanford University professor who is effectively responsible for the modern autonomous car in the shape of Stan. At Google, Thrun had already been in charge of Google's X research lab, creating Google Glass and Street View.

Just like a Stanley, Google's cars used GPS, sensors, cameras, radar, and lasers. They kitted out six Toyota Priuses and an Audi TT to drive around Mountain View, California. Then in 2011, they switched to Lexus, with a fleet of twenty-three RX450H, the big sports utility one.

In May 2014, Google showed off a prototype of its own vehicle, which really ought to have been some sort of improvement on a Lexus. Well, they called it a car, it looked rather like a Little Tikes Cozy Coupe, but not as grown up. Like a Cozy Coupe there were no brakes, no steering wheel (the Cozy Coupe one is ornamental), and no accelerator pedal, even though it could only do

25mph max. Not the Cozy. All the self driving car had was a button to turn it on and off. It also had a name and 'Baby Stan' would have been cute, instead the disappointing moniker was 'Firefly'.

Meanwhile, there were all sorts of Silicon Valley management changes within the company, but I won't bore you with the names. It is worth pointing out however that Google cars started to crash, more publically, from 2016. The point is that it wasn't always their fault and the millions of miles covered equated to a pretty decent, miles covered to accidents suffered, ratio. No one died, so it was just a standard company insurance claim and they probably lost their no claims bonus if it wasn't protected, so what was the company name?

As we all know, nothing is named to reflect what it is, who started it or what it actually does anymore. So, Google became Alphabet in 2015 and a year later the self-driving project was renamed Waymo. Appointed in 2015 John Krafcik became the CEO. Previously President and CEO of Hyundai Motor America, then after a couple of years he was starting to reveal what the grand plan was. It was ride hailing (old fashioned taxi stuff like Uber), moving around goods (like FedEx), augmenting public transportation (like buses presumably), and then making Waymo's tech available through licensing deals to other companies. He also wanted to develop what is essentially an app with some real hardware called 'driver', so there would be a system of laser-radars, sensors, and software that can then operate, or effectively drive any vehicle. That was the plan anyway.

When it comes to the whole remote driving thing, then we have to talk Tesla. If you want to read about Tesla in a teeny bit more detail, well we've hived that off to the Electrictrickery chapter, because in case you didn't know, it is electric. What you may not know is that Google had invested in Tesla. Funny that, all these Silicon Valleyers are connected and sometimes at odds with each other. Indeed, Waymo LLC v. Uber Technologies, Inc. involved former Google Car engineer Anthony Levandowski who was accused of intellectual property theft. However, the actual trial lasted just a few days in February of 2018, with Uber settling with Waymo by giving up around $245 million in company equity. Whilst we are chopping bits out of this subject to tackle elsewhere, the whole ride hailing, sharing and caring thing can be found in the Johnny Cab Chapter.

Right now, before we really dig into Tesla, this might be the ideal time to look at what counts as self-driving, as all of this could confuse a stupid person, like a consumer.

Boffins have defined five levels of autonomous driving. Each one describes the extent to which a car takes over tasks and responsibilities from its driver, and how the car and driver interact.

**Ground Zero** (a Demotorized term) is a bog standard car

with no automation at all. It is all about the driver, as it jolly well should be if it is still to be regarded as a skill. **Level 1** is a bit of driver assistance to keep the driver comfy and less stressed. Cruise Control is a bit like that and there are quite advanced systems now, which will brake and maintain distance from the vehicle in front. **Level 2** is partly automated driving, this is often to keep drivers on the straight and narrow so will brake, steer, accelerate and keep them in their lane and stop them knocking into another vehicle or object. This also includes a remote parking function. **Level 3** is highly automated driving. Handing over control from the driver to the robot, effectively a conditional system. Over long distances it is all down to the robot, but the driver must still be ready to take over and that usually occurs where there are road works. **Level 4** is fully automated driving. There are still controls and it requires a driver to take over when required. However, the point of this system is to cope with difficult situations. Oddly if the driver ignores any warning gongs and possibly Dalek like nagging, the car is supposed to park itself.

Finally, there is **Level 5**. At this point a driver is not required, even as an optional extra. There are no visible controls at all, just comfy chairs. We are being reassured that this level of robot car will be speed restricted and possibly confined to city areas.

What hardware does the robot actually use to drive itself and an ex-driver around? This stuff is worth knowing as the terms are often used as buzz phrases. Firstly we all know about GPS, global positioning system, and the clever part of the Sat Nav. In addition there is aircraft related gear such as altimeters, gyroscopes and tachometers. This is purely for navigation and has nothing to do with sniffing out hazards like pedestrians. That's why it has to work with clever things like LiDAR, which stands for light detection and ranging. It uses light (through lasers) to detect the position of objects and  mapping the surroundings in 3D. Obviously there will be cameras, front and back of the vehicle to read road signs, pay attention to traffic lights and also watch out for other vehicles in front and those behind. There will also be futuristic radar cameras. These will be next to the regular ones and won't just show where other cars are, but send out radio waves to bounce off the actual objects so they know where they are. That's useful when working with cruise control and the show off self-parking device. Making all this happen is a multi-domain controller processing all that data and information.

The idea is for the Robot to become a better driver than a human and replace them. In theory then it should be clever enough to replace a taxi driver, without the backchat, flat cap, or casual football commentary. That could send us to have a closer look at Uber, or double back to Waymo who seem to be on the same ride-hailing page. Before we do there is still the complicated Tesla issue. Many people seem to believe that their cars drive themselves already, sometimes with not very hilarious results.

Tesla was partly all about being futuristically electric and seemingly several decades ahead of all the dinosaur establishment brands like BMW and Mercedes. Then for Elon Musk's next trick it was always going to be a Tesla that saved you the hassle of ever driving again.

The software to enable an element of self-driving was included on the Model S since 2014. The company's Autopilot is an optional driver-assist system and Tesla CEO Elon Musk has repeatedly promised to turn his electric cars into fully self-driving vehicles. Indeed, Tesla claimed that Autopilot-enabled cars released since October 2016 can support full self-driving capabilities. When Autopilot is engaged, their cars can self-steer, adjust speed, detect nearby obstacles, apply brakes, and park. It was rated at Level 2 and just in case you wondered, the Enhanced Autopilot does a lot of things that other luxury barges will do. The Cruise will maintain a safe distance from the car ahead. It will keep the car in its designated lane, but the driver must be holding the wheel. It will also automatically lane change when it is safe to do so. The Tesla goes 'ping' when any car or thing gets too close to the side. Plus if the 'driver' can't parallel park, the Tesla certainly can. Then if you want to show off you can 'summon' it. That's if you can't be bothered, or can't dig your Tess out of its parking space, because there is a phone app for that.

There is a lot more of course, including blind spot detection and automatic braking and steering but let's not be an advert for the features of Tesla. Especially as they may not be perfect.

Yes, there have been crashes and the first fatal one happened in May 2016, when a Model S collided with the trailer of a truck turning left across the path of the car. Tesla said, "The high ride height of the trailer combined with its positioning across the road and the extremely rare circumstances of the impact caused the Model S to pass under the trailer, with the bottom of the trailer impacting the windshield of the Model S. Had the Model S impacted the front or rear of the trailer, even at high speed, its advanced crash safety system would likely have prevented serious injury as it has in numerous other similar incidents."

The U.S. National Highway Traffic Safety Administration (NHTSA) Office of Defects Investigations said, "Although these technologies had limitations, the ADAS [Advanced Driver Assistance System] system did not respond to an impending crash event.

"Regardless of the operational status of the Tesla's ADAS technologies, the driver was still responsible for maintaining ultimate control of the vehicle. All evidence and data gathered concluded that the driver neglected to maintain complete control of the Tesla leading up to the crash."

In 2019 the most exhaustive survey of 5,000 Tesla Model 3 owners by Bloomberg.com asked them about the automated software and how well it performed.

90% said it made them feel safer, but there were also plenty of near misses and 13% said that the Autopilot system had put them in a dangerous situation. By contrast 28% said that it had saved them from such a situation. Essentially it was regarded as a good back up, but anyone tempted to take a hands off approach to driving might well be asking for trouble. That 'Summon' feature is very much a Beta release, so not fully formed and the owners have to sort out some of the bugs. 'Incidents' when a Tesla is parking itself are at least conducted at just a few miles an hour.

These 'incidents' have not stopped Elon Musk announcing that all new Tesla models will be shipped with Autopilot capable of Level 5 autonomous driving. That's full, hands off, feet on the dashboard, couldn't care less, let the robot do all hard driving, mode. At the time of writing the final decision is down to the local authority or any regulator to approve this, but the evidence for testing artificial intelligence in the real world is not a great idea.

Yet there are self-driving cars being tested on UK roads right now. Back in 2016 the Government had a code of practice and a £20m fund for collaborative research. They wanted the UK to be at the forefront of the intelligent mobility market, which they expected to be worth £900 billion by 2025. Driverless cars were coming to London, Bristol, Coventry and Milton Keynes. The Government sees employment and pound signs, which is very nice, but will this ever become reality?

In 2019 Thatcham Research offered a set of 12 guidelines, which in their words was 'designed to minimise bumps in the road on the journey towards fully Automated Driving'. These were developed with the Association of British Insurers and related to new rules allowing Automated Driving Systems onto Motorways.

Even they said that, "The UK Government's prediction that fully automated vehicles will arrive on UK roads in 2021 is unlikely. However, early Automated Driving Systems designed only for Motorway use could be available to consumers by then. To avoid introducing a new hazard, the vehicle needs to have an effective driver monitoring system to ensure safe handover of control between driver and vehicle, and that the driver is available to take back control when needed."

The technology to make this all happen has supposed to have been happening for years. We have already mentioned that in 2015 Elon Musk, predicted his cars would be capable of "complete autonomy" by 2017. General Motors said in 2018 that it would launch a fleet of cars without steering wheels or pedals in 2019, which never happened. Also, the Ford boss Jim Hackett, said that his business "overestimated the arrival of autonomous vehicles". Then there is Chris Urmson, who was at Alphabet developing their self-driving applications who now reckons that proper self-driving cars won't arrive for 30 to 50 years. The only way to do self-driving properly is to start from scratch and build a new city and ban conventional human steered vehicles.

It is all very complicated indeed. There are ethical and moral considerations. These cars not supposed to crash, but if faced with the option of just killing a driver rather than a bus stop full of people, even if those people were proven to be terrible and just out of prison for multiple murders, what is a self driving car to do? That isn't something that a Robot should be expected to think about.

Fully driving Robots might well have been put back a bit, although not through want of trying. Unfortunately EU mandated robots are certainly coming very soon. The speed limiter is one of a range of safety features to be made mandatory from 2022, along with automated emergency braking, electronic data recorders and improved visibility built into lorries for drivers to see vulnerable cyclists and pedestrians around the vehicle. Antonio Avenoso, executive director of the European Transport Safety Council, said: "There have only been a handful of moments in the last 50 years which could be described as big leaps forward for road safety in Europe. The mandatory introduction of the seatbelt was one, and the first EU minimum crash safety standards, agreed in 1998, was another." A life saving figure of 25,000 over the first 15 years has been promised. It is an estimate and not a very accurate one.

The important point is that these systems may have to be or capable of being fitted to all brand new cars from 2022. There is no obligation to switch them on, or have them fitted. Being prevented from exceeding the speed limit may not be ideal in certain circumstances, despite there being a ‘hard-acceleration’ kick down option.

This so-called General Safety Regulation isn’t just about speed devices, manufacturers will also need to equip their vehicles with a "driver drowsiness and attention loss warning" system, which are like the posh lane departure warning features you get in many cars now which beep at you a lot.

Reversing safety with a camera or sensors will be compulsory. These are fairly old tech now, and useful if you can’t reverse and perhaps any driver having trouble doing that should have a word with themselves, go on an advanced driving course, or give up.

Rather more controversial is the "event data recorder", effectively a black box. In theory it is there to log all the relevant circumstances of an accident. It will be sold to motorists as making their lives simpler and maybe the insurance cheaper. However, this has already been called out by the ‘Ligue de défense des conducteur’s, a pro car action group, who said that once these devices are installed, the authorities can check driver data in real time, or retrospectively and quite possibly punish them for motoring offences. We will certainly see.

And finally, misleading everyone is the requirement to facilitate the installation of an "alcohol ignition interlock device". That sounds like a compulsory blow in a bag test before you can turn the ignition. Well, that sounds fairly draconian, but it may only affect drivers already subject to restrictions because of their behaviour. Otherwise vehicles will have the prep for such a device. For some, that might be just as bad as having it. It is of course the principle of thing, except that robots don't have any.

Since cars became all so very clever, there is a very real issue with hacking. Most famously some pro hackers and journalists at the computer tech magazine Wired, Charlie Miller and Chris Valasek with reporter Andy Greenberg driving a Jeep Cherokee, gave it a go in 2015. Located ten miles away they managed to mess with the air conditioning and radio.

By accessing the car's IP address it was possible to send commands to the engine and wheels through the car's internal Controller Area Network (CAN). The CAN bus carries information between the vehicle's various electronic control units (ECU) to the central controller. That handles the cruise control, electronic brakes, parking assist and control of the steering column. Although that weakness was fixed by the manufacturer, (just a fresh set of challenges for a hacker), imagine for a moment if you had really bad intentions?

There is a lot conjecture around the death of investigative journalist Michael Hastings who was driving a Mercedes C250 Coupe when it crashed and burst into flames at around 4.30am on the 18th June 2013. Counterterrorism expert Richard Clarke told The Huffington Post at the time that it was possible that

Hastings' car had been hacked.

LA Weekly revealed that in the week leading up to his death, Hastings openly expressed concerns to his friends and colleagues that his Mercedes had been tampered with. Hastings asked a neighbour to borrow her Volvo, saying that he was afraid to drive his own car.

That's a bit inconclusive, but the thing is, hacking seldom leaves any fingerprints and the fact that everything ended in a fiery ball. Tampering with a car's electronics isn't difficult, impossible, or seemingly preventable for those determined to do it. For authorities with unlimited resources, this sort of malevolent control is inevitable.

Meanwhile, artificial intelligence can still be confused by snow, flocks of seagulls or someone crossing the road with a bicycle. Despite what Hollywood might suggest, Robots never will be able to replicate a human being. We really are too complicated, fickle and clever. It takes us just a few hours to learn to drive and some might say it shows the very real shortfalls in the driving test. However, after millions of hours and miles there still isn't a robot driver clever enough to deal with the cut and thrust of rush hour traffic on a rainy late November afternoon, just outside of Chorley.

Until conclusively proven otherwise, best not to trust a robot to do the driving for you.

*Electric Milk Shake*

## 14. Electrickery - *Like milk floats, faster, but not as useful and less clinky.*

An electric car is the future, right? Not just cars because Electric Vehicles (EVs), includes lorries, vans, (there are already trains) and planes are all part of our bright new zero emission future.

The funny thing is back at the dawn of motoring, electricity was just one of several propulsion methods jostling for position. Incredibly, and in the face of all the evidence, electric vehicles were not invented a couple of years ago by Elon Musk.

If anyone has a claim on early EVs as part of their heritage then it is Porsche in 1898. That's when Ferdinand Porsche's design was dubbed the Egger-Lohner electric vehicle, the C.2 Phaeton model, or even the P1 for short. Take that McLaren. The P1 was so good it won a Berlin Road Race in 1899 for electric vehicles. Then to prove that there is absolutely nothing brand new anymore, there was the Austrian Lohner-Porsche from 1900 where electric motors were in the wheel hubs and charged by a couple of on-board petrol engines. There's the first hybrid car. So when manufacturers tell you they have invented something new, they most likely haven't. It has all been done. Centuries ago.

Here's the absolute truth the, EVs were once an integral part of sustainable British life. People went to work on things called trams and trolleybuses and their early morning pint of milk arrived via the largest fleet of EVs in the world, better known as milk floats. They were quiet, made a low drone and a reassuring clinky noise as the bottles tapped together. Yes, they were glass bottles, because the old ones were collected by the float, taken back to the depot, cleaned and reused. Even the foil lids could be kept, put into an envelope and sent to the BBC children's programme Blue Peter who would turn them into money for charity. That was over 40 years ago.

Right now plugging an EV into a wall socket and waiting for it to have enough electric juice for a trip to the shops to buy a pinta is seen as progress and indeed the sustainable, green future. Even if it is something of a compromise. But what did the Future look like in the nineteenth century?

Electric Vehicles were actually horse drawn carriages without the horse, just like the petrol and steam driven ones. Who first decided to use batteries is a matter of national pride. If you are Hungarian then the date 1828 might be tattooed on your heart, because that's when Anyos Jedlik made a battery and stuck it in a model carriage. Much more convincing though was Scotsman, Robert Anderson who actually took a full scale carriage and at some time between 1832 and 1839 installed non-rechargeable batteries and drove people around in it.

At the same time there were pioneers in America, Holland and Wolverhampton. The most remarkable story relates to a genuine motoring hero and inventor Thomas Parker. In 1884 he built an EV, which he used, in and around the West Midlands. That's a decade before the petrol-powered car arrived in Britain and also a year before Karl Benz first patented his first full production carriage. Apparently, Parker built a second car and it was being shipped to Paris for evaluation and manufacture when the boat carrying it sank in the English Channel.

EVs became a biggish thing. There were a small amount of cabs running around in New York and EVs were marketed at well-off women because there was no brutish starting handle to deal with. EVs were certainly fast cars. In 1899 Belgian Camile Jenatzy set a new land speed record at 65.79 mph. Oddly enough that record was beaten in 1902 by Leon Serpollet's steam car, which raised the record to 75.06mph.

So what killed off the battery car and the steam car for that matter? Batteries were penalised by weight and range. Pretty much the same as today, but much worse. Steam cars were great, efficient, relatively clean, although they did require a small bonfire to be on board, were fairly complicated and didn't exactly fire up at the hearty swing of a starting handle. They took time to come to the boil.

Henry Ford gets the most blame for coming up with the fabulously successful Model T in 1908. Oddly a big fan of biofuel, his cars had adjustable carburetors that could run on the stuff, so that farmers could not only distil their own hooch, but four star as well. Making a cheap, efficient and reliable car had everything to do with it and the fact that the Model T needed petrol only encouraged the electric downfall. There isn't a huge amount of evidence that the oil companies actively worked against electric cars. However, J.D. Rockerfeller through his Standard Oil conglomerate certainly bribed and bought out suppliers to monopolise his marketplace.

In Britain there is a long history of electric commercial vehicles. Never mind the milk floats, department stores and other delivery vans operated for decades in London, electric dustcarts in Birmingham, various EVs were used for local bread deliveries as well. Changing consumer habits meant more people were at work, doing bigger weekly shops from supermarkets. Indeed, traditional EVs were replaced on some rounds by diesel vans that got through starter motors every couple of months and were completely unsuited to the repetitive stop/start nature of that work.

There was certainly considerable discussion of the potential for electric vehicles and government provided some funding for the development of sodium sulphur batteries for road vehicles. The Central Policy Review Staff of the Cabinet Office published a report at the end of June 1974 recommending the promotion of electric vehicles as part of an energy conservation strategy. Professor Colin Buchanan and various government ministers all made positive speeches at a meeting of the

Electric Vehicle Association without anything significant happening as a result.

There were electrical experiments throughout the '70s in response to the fuel crisis, most of them concerned with public transport. This didn't move the whole mass movement of commuters onwards in a quiet and orderly, three-pin plug type, fashion. Most local authorities had enough to do, simply making sure that the corporation diesel buses would sort of run on time. Essentially no one was into establishing electricity as an alternative, or had enough of a budget to make a difference.

My old friends in the Battery Vehicle Society had long lists of experiments conducted during these busy times and as recently as 1993 and 1995 there was a bus trials in Oxford. Electric buses were restricted to the city centre but technical problems and lack of funds meant they were discontinued and the comparison with diesel buses ended quickly with no concrete conclusions.

Apart from the ubiquitous milk float, probably the best-known British electric vehicle was the Enfield E8000 a.k.a. E8000 ECC or even Electric City Car, but not actually built in the north London suburb it shared a name with. The company was started by a Greek millionaire Giannis Goulandris and based on the Isle of Wight.

The Enfield 8000 was the result of a competition run by the Electricity Council in 1966 to find a new design for an electric car.

Introduced in 1973, 120 examples were built and 65 were used by the Electricity Council and electricity boards in the south of England. There were aluminum body panels, with suspension parts from a Hillman Imp, adapted mini doors and a rear axle from a Reliant three-wheeler. It was pushed along by an 8bhp electric motor, fed by eight 12V lead-acid batteries. It would do a few miles an hour short of 50mph, and could go for 50+ miles on a charge. It did though have a basic but clever kinetic energy recovery system, recharging the batteries fully while freewheeling downhill.

Most were Electricity Council company cars for regional managers who probably weren't flogging enough watts each week. Retail sales though were not very impressive at all. It wasn't a looker, resembling a '70s kit car and most buyers would rather have the constituent parts from the donor cars (including the Reliant) rather than the actual Enfield. The price was a stiff £2808 at least double that of a mini which had more room, would do a good few hundred miles on a tank and be a proper hoot to own and drive. Just as today, 2020, motorists won't buy an EV because the range is poor or at least worrisome and they are at least twice the price of a conventionally powered car. Consumers aren't stupid, that's why they have to be bribed or forced to adopt a new technology.

Manufacturers rather liked the whole idea of simply plugging one of their models into the mains. It certainly made sense to cut their ties with the oil companies. After all, they were convinced that the black stuff would have run out by the 1990s anyway. If car makers wanted to stay in business, they needed to learn to adapt. Over the years there were lots of experiments, concepts and clever model names with E in them, which I won't bore you with.

That was then, let's silently fast forward to the real second coming of the electric car when a big manufacturer decided to go out on an electrification limb.

The brave, cutting edge company taking a chance was General Motors (G.M.). In 1996 they launched the first commercially available electric car in America. It was appropriately enough badged as the EV1. You could plug it in at home or find one of GM's battery parks to give it a boost. The EV1 was leased and became the vehicle of choice for Hollywood celebrities (well Ed Begley Junior had one, if that counts). It also looked very cute. Rather than going for the easy option and sticking an electric motor in a standard model, this one looked like it really was driving into the future. Blunt fronted, because you don't need a radiator, curvy and rocket shaped it resembled a miniature 1950s Sci Fi rocket, but with wheels. Buck Rogers would have driven one in America, Dan Dare in Blighty.

Here was a car perfectly designed to fit in with the dark times. California was still fighting against smog and the state regulations specified that two percent of all new cars sold had to be electric by 1998 and 10 percent by 2003. That was the dream anyway.

Just over 1,000 EV1s were produced by G.M. before the company pulled the 'plug' in 2002, due to lack of demand. It might have looked the part and been futuristically retro, but they didn't cut it in the showroom. Even other manufacturers, who took the conventional route of electrifying existing models, didn't exactly set sales records. For instance there was a Toyota RAV4 and also a Honda EV Plus, but handfuls were sold. Car manufacturers also seemed to resent being told what to build by the California legislators.

Despite ploughing $1 billion into the EV1 project, GM took them back when the lease expired and broke them up, which is something they were perfectly entitled to do. EV fans say that GM killed their car and indeed, there was a very good film made with that plot, "Who Killed the Electric Car?" in 2006. The evidence is there just were not enough real world motorists willing to sign up for a small battery car.

The third coming of the electric car occurred a year after the EV 1 but it wasn't fully electric. What Toyota proved was that if you are going electric, don't go the whole maximum volt hog. Not if you want to sell a lot of cars, plus get the goodwill and alternative, low emission love, then all you needed to do was build the world's first mass-produced petrol-electric hybrid vehicle, badged as the Prius, (Latin for 'to go before').

At the heart of the new car and every one since was the revolutionary Toyota Hybrid System. This was an integrated package, which could be conveniently inserted under the bonnet. Providing the main motive power was a highly efficient four-cylinder 1.5-litre petrol engine. It was connected to a compact and high-torque electric drive motor, and a separate generator. The final component was a powerful nickel-metal hydride battery pack behind the back seats. The exterior was unremarkable, but had a low drag coefficient of Cd 0.29 to make it slippery and more economical with a theoretical driving range of 560 miles. The combined fuel economy was 57.6mpg and the power plant produced just 114g/km of CO2. Clean by the standards of the time and very frugal too, effectively a diesel without the clatter, the smoke, or social embarrassment.

Three years after Japan, the Prius finally came to the UK in October 2000 and it was great that the company realised that cynical Brits would need some convincing to adopt the new technology and reassurance about battery operation and life. With that in mind, Europe's first five-year mechanical warranty was introduced and there was a leasing scheme called Prius One to protect buyers from any future residual value risk. Not only that, Toyota committed itself to handling the recycling at the end of the car's useful life. No reasons not to buy then.

Judged purely as an efficient, reliable and practical car,

the Prius made all sorts of commonsense as a family car. Eventually there were alternatives, the Honda Insight that was like a prettier and better resolved EV-1. Otherwise Honda offered their dreary Civics with Hybrid propulsion.

Then there was the Chevrolet Bolt, which was badged in Europe as the Vauxhall/Opel Ampera from 2012. Technically not a hybrid although it did have a couple of engines. A clever thing, if wasteful of weight and technology in that what you have is an electric car without the range anxiety thing. It would run on just the electric motor for between 25 and 50 miles. Once the battery is run down the 1.4-litre petrol engine wakes up and effectively acts as a generator to provide an additional 310-mile range. The electric system had an 8 year/100,000 mile warranty which showed real confidence in those lithium-ion batteries. Trouble is, this was not a cheap set of wheels at all, £33,750, less the taxpayer grant. The miles per gallon looks out of these world, at around 250mpg+. According to green transport experts what you are supposed to do is maximise the electric range which is up to 50 miles so that the actual emissions amount to 69 g/km. Factor in the extra petrol assisted 310 miles and the actual mpg amounts to around 40. The bottom line is that you don't have to visit the petrol station that often, which is a huge bonus.

This is the perfect electric car for cynics who can afford it. You can't help thinking that you are carting around a generator, and that's exactly what one version of the BMW i3 and i8 did as well. Because it was BMW and looked like nothing on earth, they managed to get away with it. Oddly enough, sales were not spectacular and neither of those models is coming back, but the electric tech will be spread around the more conventional BMW Series models.

The Fourth coming of the electric car? Well that would be Elon Musk's Tesla. The thing is though there was supposed to be a real Tesla car back in 1931, but it seems that the whole story could be a load of made up nonsense, sadly.

Inventor Nikola Tesla took a fantastically expensive Pierce-Arrow without an engine, but all the other running gear and installed an 80-H.P. 1800 r.p.m. electric motor, which was hooked up to the transmission. Legend has it that Mr. Tesla simply went into a local Maplin, which didn't even exist back then, just like now and bought some radio valves and resistors. He put together a circuit in a wooden box, placed it on the passenger seat and connected it to the electric motor. Electricity is all around us and Tesla was clever enough to harness it and direct it to the motor and hit speeds of 90mph in silence. So this was like picking up radio waves, but instead of Chris Evans it was proper power, which originated, and I quote from Mr. Tesla himself, "From the ether all around us".

There are no blueprints or clues as to how this all worked, but here is the dream for so many, wirelessly powering everything from a toothbrush to a Jacuzzi. This is how we would all want power to actually work, by being clean, invisible and for the environmentalist in all of us, produce no emissions. This may well be real and perhaps those nasty power companies all clubbed to together and are sitting on the plans, or simply threw burnt them in one of their coal fired power stations.

Finally, we come to Elon Musk who seems to have reinvented the electric car for the current cooler, but no so clever generation. Probably the most remarkable thing about Elon Musk is that he is a South African, Canadian, and American. These were the stops along the way to American citizenship and success. Behind it is all is a very complicated story and we are not interested in the spacecraft, that Boring company, or his beginnings as a humble payment processor, just the cars. Except that he didn't start Tesla Motors, Martin Eberhard and Marc Tarpenning did in 2003.

Musk turned up in 2004 investing $30 million to become the chairman of its Board of Directors. Tesla Motors was all about building all-electric, super green transportation. The added twist that Musk introduced was he would also develop these into affordable autonomous vehicles. We covered Robot Cars a chapter ago, so you can disappear down that Level 3 automated rabbit hole. Musk also attracted further investment from the Google's Larry Page. Proving that everyone is connected in Silicon Valley.

The first Tesla, the 2009 Coupe was an interesting, but

not very affordable. A small Lotus alike sports car would be a dead end for a car manufacturer because there is a limited market for models like these. Better to aim for a bigger and more profitable market, hence the Model S from 2012, which was a luxury executive saloon. It could travel up to 300 miles on a single charge. Tesla also opened its first freestanding charging stations, called Superchargers. It began with six located in California and has expanded to operate more than 1,000 worldwide. These locations offered free charging to Tesla owners, much faster than using a common wall socket outlet.

On the face of it, Tesla business looked good when in 2013 they actually made a quarterly profit and had plans to build a so-called Gigafactory making the batteries. Beneath the surface though, it all wasn't that great. Build quality was all over the place and even Musk said that the cars 'sucked'. There was certainly quite a big cash flow issue and for at time Google were lined up to buy Tesla for $6bn. A concerted sales push meant Tesla got enough money to tide them over, but this wasn't exactly a normal business. Not one that had to make products and profits and boring stuff like that, Tesla was and still is, a religion.

Owning a Tesla was a making a statement. It marked you out as cleverer than everyone else. It was also reassuringly expensive too. Whereas the Toyota Prius was a bit too affordable, it didn't make the elite feel as special as they thought they were. OK so there were free chargers at home and on the road, but the starting price back in 2013 was £72,850. The thing is though the British Government, like many others, were quite happy to dole out a £5000 subsidy to buyers of these and other electric cars who could easily afford them anyway. If something is any good, a subsidy is never needed. Politicians love electric cars, because they believe EVs are zero emissions and so to make a point, they readily give away money that isn't theirs.

Instead of buying a Mercedes S-Class, Audi A8, BMW 7-Series or Jaguar XJ, a Tesla was the luxury large saloon of choice for the environmental show offs. The thing is there was nothing to directly compare the Tesla with, so as a luxury 'lecky' barge, it was in a league of its own. Maybe the company should have concentrated on the quality and door handles rather than getting into areas such as making houses, which were not directly related to making their cars, better.

Related to all this, Tesla had changed its name by 2017 from "Tesla Motors" to "Tesla, Inc." to indicate that the new scope of its energy products. Rather than concentrating on making cars properly they were now all about energy solutions.

The people's Tesla did sort of arrive in the shape of the Model 3 in 2016. Prices started at £42K and were pegged back to £37,340 in 2019. It is safe, quick and had groovy roof hinged doors, so it does look as though it comes from the future. Musk predicted that they would make 200K of them by 2017, but that never happened.

It does seem that buying, owning, investing in and running a Tesla seems to have been a constant disappointment.

There were regular production shortfalls, in 2018 it took more than three months to make 2,500 cars when it was supposed to be 5,000 a week, because of failing supply chains. There were financial embarrassments and eventually legal ramifications when Musk said he planned to take the company private "at $420" and didn't. The details are dreary and irrelevant here, but there was a £20m fine and Musk stepped down as director, but remained CEO.

Just before Demotorized came out, the Tesla Cyber Truck certainly did. It is a pointless, sharp edged bundle of nonsense. Yet another SUV without any real purpose, it demonstrated perfectly how Tesla distracts everyone from more pressing company issues, be it the bizarre financial big dipper ride it is on, or the quality of goods on sale. Isn't Elon Musk clever? Well, he had better brace himself for what is coming next.

Tesla no longer has it all their own way as the top end of the electromarketplace. The Europeans are plugging into the profitable end of the market. It's not just because Audi and Porsche can build proper cars, but the stringent EU Co2 regs have been designed to force mass battery adoption by at least 2025.

"Electric cars in general won't be as profitable as cars with combustion engines," BMW CEO Harald Krüger said in 2019. Meanwhile, Daimler has drawn up a €4 billion cost-cutting plan to offset lower profitability from its electric cars. Honda CEO Takahiro Hachigō, spoke to the Financial Times and said that that profitability was dependent on them becoming mass scale models. Making the chassis and drivetrain for EVs is significantly cheaper than for internal combustion engines, so the prices of cars will come under sustained downward pressure. However, EVs have lower profit margins, and are more reliable so will last longer. Tesla has been clever and concentrated on the upper end of the market. Plus the really expensive battery technology has been kept in house, which means that long term they may start to make some serious money. Right now though, going electric seems rather optimistic.

Just say that overnight it was possible to switch from filthy fossil fuel to electric vehicles in say the world's largest car market like the United States. The amount of effort and mining required would be mind blowing. A back of a recyclable envelope, calculation is very sobering. It would quickly reveal that the world's current Cobalt output as a by-product of nickel and copper mining activities, would need to be increased by a mere eighteen times to cope. Not only that, working around cobalt can be dangerous and exposure to it can cause heart, vision problems, and more. It gets worse, because cobalt mining is not unconnected to child labor in the Congo as they struggle with buckets of spoil, rather than being at school.

Then there is Neodymium a rare-earth element. It is a hard, slightly malleable silvery metal that quickly tarnishes in air and moisture so doesn't look that pretty, or useful once mined, but don't worry you will need nine times more of that. Oh yes it is also very irritating to the eyes and skin once dug up, but very good at being a magnet. You would be able to pick up a pin you dropped ten years ago in your back garden in Swindon and now you live in Crewe. Yes, you need nine times more of that than is mined at present.

Lithium is needed for batteries, but not the salt version used in psychiatric medication. It might come in handy though if you are a bit down about the whole battery range anxiety thing. Anyway, Lithium Ion is good because, among other things, lithium ions and electrons move easily back into negative electrodes making batteries rechargeable. You'll need at least seven times more lithium than was previously mined. This particular raw material also happens to be a limited natural resource, so prices are on the rise. Also, when it comes to rare minerals these include praseodymium, lanthanum, dysprosium, and neodymium, which are getting rarer. Your future electric car will become ever more expensive and exclusive.

Oh yes, and there's more. More copper, four times more copper.

Essentially, a lot of chemical elements are needed and that means plenty of environmentally disruptive mining, yet there is an on-going fuss about extracting fossil fuels.

Once the batteries run flat, or don't charge as well as they did, what happens next? Both lithium-ion and Ni-Mh batteries can be recycled. However, the market for these is currently not that large. Old lead acid batteries, the sort that you found in milk floats and still have in normal cars for the wipers and horn, are actually very recyclable and 96% of them are reused.

When it comes to electric car battery packs, Nissan partnered with power management firm Eaton for its car batteries to be re-used for home energy storage as they can still have up to 70% of their capacity when they stop being good enough to power electric vehicles. However, recycling isn't easy. Commercial smelting processes can recover many metals, but lithium, which ends up in a mixed by product, is only extractable at a price. All this is important because car manufacturers will be responsible for their cars at the end of their lives and many need to put a recycling plan in place, like Toyota did with the Prius.

Electric cars might be simpler, but they still get safety recalls. Volkswagen, Porsche, and Audi recalled their hybrid and electric cars due to fears about poisonous cadmium. There is also the risk of electrolyte spillage or leaks. The last thing you want though is a fire, because potentially they can burn for days. Lithium-ion batteries for instance, even after the flames have been extinguished, still contain energy and keep sparking, reigniting the car. Water, used to put out fires, can cause short circuits, leading to more sparking. Foam or power-based fire extinguishing material is better for electric car fires.

Range anxiety is supposed to be idiotic say EV fans, because for 99% of the time we only travel just three miles. That's a run to the shops and back, so we could all easily pay the minimal £26K required for a new EV and scrap our dinosaur era petrols and diesels. Except that life isn't really like that and there are basic journeys such as visiting relatives, or getting to a business meeting that are still out of an EV's range.

There is a difference between a top up and a full charge, but it may take much longer than scoffing a coffee and a bun. Things may well change and technology will improve exponentially, but there is nothing to gain absolutely everything to lose, mostly money, by being an early adopter of just about anything. This is worth constantly restating, if something is life changingly brilliant, consumers will go and bag one for themselves. They don't need to be pestered, offered incentives or prosecuted for not doing so. Make EVs brilliant and buyers will come and then go silently on their way.

EVs depreciate at a faster rate due to fact that they are alternative technology and the car trade traditionally views them with suspicion and the battery life will inevitably degrade over time and are the most expensive part of the vehicle. Best not to own an EV, just lease it, return at the end of the payment period and start again. These don't seem like cars to own or cherish because of the technology. Even so they should be cheaper to run as there are less moving parts, so in theory a service should be cheaper. Operating upgrades when they come will be through software patches and Tesla has been known to ration those. You could be charged a fortune to plug into a dealer diagnostic system, but what alternative would you have? Most EVs have a separate warranty for their battery, and usually it runs for longer than the car's warranty which can be between five and eight-years. Battery life is still a worry and that's the reason why buying a used EV is always going to be problematic, unless there is a decent warranty.

Batteries are the beginning and end of the issue, if you can't get the batteries then the cars won't be sold. Waiting lists have been an issue with Tesla from day one. It isn't just about popularity, it has everything to do with supply. They are not just building cars, that is the easy bit, what they haven't been used to doing is building batteries. That has caused a battery bottleneck. In 2019. Clean Technica reported: carmakers can't match production with demand. As a specific example in the Brussels Times, Audi lowered the 2019 production forecast for its Belgian plant where it builds the electric e-tron, by more than 10,000 cars to 45,242 units. Audi won't comment on whether it is battery supply, but they also put back their next electric Sportback to 2020.

Presumably Volkswagen have everything carefully in

place for their all electric ID. Models, note the artfully placed full stop. As I type this sentence Ford have brought out a new Mustang, which is an SUV and also happens to be electric. You either view that as stunning and brave because there is another Useless Sports Versatility thingee in the world, or the trashing of a once proud model name. Then again, it is their name and they can do what they like with it, but give the world ten minutes and yet another EV is launched on an unsuspecting world. Great news if you are in the battery business. However, some people can see the actual business battery reality.

Indeed, James Dyson has freed up some valuable battery capacity. "The Dyson Automotive team have developed a fantastic car; they have been ingenious in their approach while remaining faithful to our philosophies. However, though we have tried very hard throughout the development process, we simply cannot make it commercially viable."

There you go, better to stick to premium priced hair dryers and vacuums, don't mention the washing machines, than get into the murky world of marketing motors. Dyson is a clever bloke, that's why he is a billionaire and chasing down this impossible dream of zero emissions nirvana is too big a job for even him. Not worth the effort and significantly it isn't going to bring much in the way of profits. He doesn't have to stay in the car business if he doesn't want to, although he could probably make a really, really good electric scooter. A little more on that in a few pages time.

Luckily EVs are still at the expensive gimmick stage. Motorists who are richer and want to show off can spend their money how they like. What you need to worry about is when every car is mandated to become electric. Then the cost of electricity will rise even further, possibly to the overtaxed petrol/diesel levels we have now. When there are no alternatives then everyone will ultimately suffer.

Because of the Euro CO2 regulations motorists may never get a real choice over which cars to buy. Indeed, carmakers may have to ration petrol car sales in order to meet those 2021 targets, which specify that group CO2 emissions must not surpass 95g/km, or be liable for multi-million-euro fines.

Fiat Chrysler Automobiles (FCA) and Tesla have been working up a carbon-swapping scheme as Tesla gets a much needed cash injection and Fiat dodge the fines. That's why it seems like every manufacturer is piling into the EV and Hybrid marketplace, because they have to.

Obviously it is all about money. The fine is effectively €95 for each 1g over the carbon target and then that amount is multiplied by the number of vehicles sold, or built per year by that manufacturer. Most car makers build at least 200,000 to 500,000+ annually. That's quite a big figure, even if the carbon amount is modest. Then that money is put in some suitcases and handed over to the European Union and heaven knows what they will do with it. So billions of Euros are at stake here and if you want to buy a practical, cost effective diesel or petrol vehicle, you are being denied that sensible choice.

When it comes to stupid rules there will always some unintended and equally daft consequences. What is slowly happening is that the City car is being killed off. You know those small, frugal, cheap and practical vehicles that used to live in town and city centres, getting dented and going about their daily business of shopping, commuting and stuff. If you believe in low CO2 output, then what would be better than a tiny hatchback? They were cheap and insanely cheerful. Often they could crack the holy grail of being a sub 100g/km by a decent margin. To get any lower, or at least a mythical zero, that would mean turning electric. It means shoving in batteries so they weigh literally a couple of tons and rather than being affordable, they can't be anymore. When Demotorized came out a basic now old model Vauxhall Corsa was around £10,810, the new E for electric one is £26,490. The Taxpayer does chip in with a £3500 grant, but it is literally nuts to pay so much for so little.

What consumers are now being told is just plain stupid. That one of those giant electric Sports Utility Vehicles is just as environmentally friendly and economical as an electric City Car. Not worth unpicking all that is it? Buyers with the money are going to spend money on something big and show offy. That is the future and that is what consumers are going to be offered.

A more sinister aspect is that once the roads are filled with electric cars, every vehicle will be remotely controllable, see the Robot Cars chapter. Every journey can be monitored and presumably can be turned off if need be.

As the rapidly depleting power grid cannot supply the number of cars on the roads today, millions of people will lose the use of a car, and be restricted to using a not necessarily adequate public transport system. Driving will then become the prerogative of the rich. By all means apply for a permit to use an electric car, but that is going to cost. Welcome to the George Orwell 1984 School of Motoring.

Never mind, can't afford one of those fancy EVs? Tesla a bit too rich for you? Don't think that the Porsche Taycan is right for your image, or maybe it has just too many seats? Don't worry, car manufacturers have got exactly what you want. Scooters. Not just that, it is a scooter with a battery. Seriously.

There are a few to choose from, and first to be popped onto its inadequate centre stand and tip over, is from SEAT. The Spanish operating arm of Volkswagen came up with something called SEAT Urban Mobility and very proudly showed off their e-Kickscooter concept in 2019. Hyundai had its own Ioniq e-scooter concept in 2017, but things have gone a bit quiet on that. However,

Peugeot has offered the Micro e-Kick scooter as an accessory for the 3008 GT SUV though no one has ever seen one and the folding BMW X2City e-scooter is also available for a little over £2000. It has usefully big wheels for bumping over pavements and getting in people's way.

Oh yes and Ford, bought out Spin. This is a bicycle- and scooter-sharing company launching as a dock-less bicycle-sharing system controlled by a mobile app for reservations.

So have car companies, gone mad, hip, or are they seeing the bigger picture? Well. These E-scooters can be labeled by pretentious people who do Power Point presentations as 'micro-mobility' solutions. Effectively E-scooters deliver the quick and easy last-mile of travel in the urban landscape that we are all supposed to occupy in the near future. Presumably these little scoots might help their carbon footprint, but it must also be about manufacturers not seeing big paydays from selling EVs.

The Peugeot Group's electric and connected boss Helen Lees said, "EVs are far simpler (than internal combustion engined vehicles)," she explained. "They need less parts, less time in the workshop. Ultimately, it means less time in aftersales. That's why we've chosen to diversify into areas such as shared mobility."

Unfortunately EVs are causing job losses. Yet the EU had suggested when releasing their CO2 targets that there would be 60,000 jobs created by the new technology. The reality is somewhat different. At the end of 2019 Audi announced cuts of 9,500 of its 61,000 jobs in Germany by 2025 to make more money available for electric vehicles. They said that the cuts - which aim to save €6bn will be achieved through early retirement. Daimler also announced that they would cut more than a 1000 jobs by 2022. Indeed, there was a lot of restructuring, code word for redundancies among all the world's top car makers. The sheer expense of building cars people can't as easily afford and don't need, unless they are forced to have them, obviously isn't going well.

**Green Plates Good?** As Demotorized is published there is a debate about whether EVs should have Green Number plates. The government has asked for public input on the proposal. What we possibly have here is the creation of a two-tier car ownership system. Those who can afford an electric car and those that can't. What about Hybrid vehicles, which may be able to run as a full EV around town, do they get a half green plate? Or does it just turn green when electricity is used? As usual with these types of daft show off proposals, there are more questions than answers. Electric Vehicle owners could get reduced parking fees, have designated road space or be allowed into bus lanes, at least until they clog those up.

Just a reminder before this chapter ends that politicians, love electric cars. They are 'Zero' emissions.

Quiet. Controllable. Not just by using the accelerator, but electricity itself can be switched on and off. Also, once these vehicles are connected to each other, the government will know where everyone is. Oh and be able to switch EVs on and off.

As mentioned previously, EVs have needed tax payer subsidies to get initial sales traction and when that stops, so do the sales. Fortune Magazine reported in July 2019 that electric-car sales fell for the first time when China scaled back purchase subsidies. Monthly sales worldwide fell 14% to about 128,000. Sales also fell in the UK at the same time as alternative fuels – including battery electric vehicles and hybrids – were down 11.8% in June 2019 compared to June 2018.

The Society for Motor Manufacturers and Traders blamed the downgrade of government subsidies in October 2018. The £2,500 subsidy for plug-in hybrid vehicles was cut altogether, and the government reduced the grant for pure electric vehicles from £4,500 to £3,500.

It is important to remember that there are propulsion alternatives, which are not electric at all. Hydrogen fuel-cell vehicles emit nothing but water vapour, have a long range and can be filled almost as quickly as a petrol or diesel car. However, making the hydrogen is expensive and delivering it to vehicles requires an infrastructure that barely exists. This is a complicated issue but all governments and most manufacturers are betting on giant AA batteries.

EVs do a couple of things quite well. Firstly they have the ability to move pollution that they do create to somewhere else. Be it the production of the vehicle itself, or the supply of the power, which could be created in an environmentally acceptable way (renewable, which does have an environmental cost ultimately) or even unacceptable way (nuclear, except that it is a sustainable and clean).

Want to be environmentally neutral? Walk to work. Walk to school. Walk to the shops. Or use one of them bicycle things. After all, that's what people used to do. Otherwise buy and run and old car. See Bangernomics.

*Uberfication*

**15. Johnny Cabs are Real** - *Ride hailing, sharing, and caring non-car owners.*

Total Recall was a magnificent piece of Schwarzenegger science fiction nonsense. One of the best bits was when inside the cheese wedged shaped self-driving cab, Arnie ripped the irritating Johnny's head and body off the joystick and took over. Not sure that is possible with what is proposed with the coming generation of pilotless cars without a steering wheel, or pedal to their name.

At the moment, there is still a taxi driver to get annoyed with when they take the long way around the one way system. Minicabs are part of transport landscape, but there does seem to be an agenda to get us out of owning, or driving a car and in the back of a hybrid powered app. That app, is Uber.

If you don't know, or haven't used Uber already, the smartphone app allows users to hail a minicab at the click of a button. It then sends a named nearby driver to their exact location using GPS, calculates the cost of the subsequent journey using an algorithm and collects payment via the user's credit card.

In just one paragraph you can understand why it is so easy, accessible and popular. It is also cheap. Uber is between 30 and 40 per cent cheaper than the usual black cabs in major cities.

Specifically in London, Uber drivers are unqualified,

have never done 'The Knowledge' of London's streets for two years on a rented moped in their spare time, whilst working double shifts in a supermarket. No, instead of plugging in their brains, they plug in a satnav.

Uber kicked off in 2012 in London and by 2015 there were more Uber drivers than black cabbies, which amounted to a fairly modest 24,000. The number of minicabs on the streets was a massive 89,000. There were concerns about the safety of passengers when Uber drivers didn't need to pass any tests or undergo background checks. In 2015 it sadly came as no surprise when Samson Haile, from Brentford in West London, was jailed for eight months for sexually assaulting a woman in his vehicle. It must be remembered that the worst case of sexual assault and rape was black cab driver John Warboys with more than 100 victims. He was imprisoned in 2009.

Nevertheless, the police were finding that there was a assault allegation being made against Uber drivers every other week. Mayor Boris Johnson was behind a set of new rules called Private Hire Regulations Review limiting the total number of mini-cabs, requiring drivers to pass a written English test and insisting that all private hire firms should wait at least five minutes between accepting a booking and picking up a customer.

Uber obviously objected and launched a petition. This triggered the Downing Street office to get involved and that included Chancellor George Osborne and Prime

Minister David Cameron. It included the offices of Cabinet ministers, such as Business Secretary Sajid Javid. According to the Daily Mail, they were effectively lobbying for the business interests of Uber. That may or may not have led to a turnaround in January 2016 when most of the proposals were dropped.

At the Licensed Taxi Drivers' Association, 'The outcome of this so-called review was decided by Downing Street, who behaved like paid-up lobbyists for Uber,' said its leader Steve McNamara at the time.

It is possible to explain all this by making a lot of obvious connections courtesy of the Mail Online. PR Executive Rachel Whetstone worked with fellow Public Relations operative, David Cameron at the same firm, media company Carlton. They also worked together along with George Osborne at Conservative HQ in the '90s. Whetstone was also the wife of Steve Hilton, Cameron's original Downing Street Director of Strategy. Leaving politics in 2005, she joined Google as vice-president of global communications, moving onto Uber in 2005.

In his 2015 Budget, the then Chancellor, Osbourne, said that as a cost cutting measure Government staff would be required to use car-sharing apps. Basically Uber.

This was just the tip of Uber connection iceberg and the last big one would be ex Chancellor Osborne who took a job at the U.S. investment firm Black Rock who had invested a reported £124 million in Uber in 2014. Uber was always an irresistible force, but it certainly received plenty of help along the way.

According to the Public Carriage Office there are now fewer Black Cabs at 21,000 and more than 120,000 mini cabs, although Uber only ever talks about drivers, with the number standing at 45,000. Cab sharing is a very big thing.

What London and the other city centres (Cambridge, Birmingham, Leicester, Nottingham, Stoke, Leeds, Newcastle, Merseyside, Sheffield, York, Belfast, Glasgow, Edinburgh, Cardiff, Brighton, Bristol, and Portsmouth) have ended up with increasing congestion and air pollution which is not what their leaders would want. Lots of illegal parking, and accidents don't help but at least they are carefully documented by the 'Black Cab' professionals.

It is a free market and there is more competition, but it is not amongst equals. This is an American app company damaging the infrastructure of Cities and having a downward pressure on the income of highly qualified indigenous Black Taxi and minicab drivers. The way Uber operates has also ruined the business model of so many well-established independent mini cab companies who cannot possibly compete. In 2015, Uber paid just £411,000 in corporation tax and this has led to the Inland Revenue investigation. The company reportedly owes more than £1bn in unpaid UK tax. One anomaly is that Uber does not pay VAT on fares because the company says it is just a middleman between its self-employed drivers and passengers. Yet this is a company worth $50.3 billion according to the New York Stock Exchange, but has never made a profit.

Certainly Uber finances are very creative indeed using the 'double Dutch' structure, where all fares paid by British customers are funnelled to a firm in Holland called Uber BV. Then most of it gets pinged back to the drivers and what is left, which will be billions ends its journey at Uber International CV, which is located in the offices of a Bermuda based firm of lawyers.

Never mind, surely all those drivers, 60,000, at the last count are pumping loads of income tax into the economy. Actually most are reliant on tax credits and other benefits. They are classified as self-employed despite a case brought by nineteen British drivers. They successfully argued at an employment tribunal that they should be regarded as employees, rather than contractors, and paid the 'national living wage', get sick, holiday pay and overtime. The drivers said they earned as little as £5 an hour. The court did find in the drivers favour, but Uber were appealing and they still seem to be getting their way.

Being an Uber Driver is complicated, or simply expensive, as there are lots of things to think about. Buying or renting a car, then fuelling and maintaining it. That would seem to cancel out much of the actual income from being a taxi driver. The average amount according to the uberdriverlondon.co.uk blog it is no more than £10 to £15 an hour. Then again, the behaviour of the drivers hasn't been great.

A serious tax investigation may be the least of Uber's worries as the company has been fighting to keep its licence in what is clearly the City where they turn over the most revenue that they can then process very creatively.

Towards the end of 2019 Uber was been stripped of its operating licence by Transport for London (TfL) over passenger safety fears. A change to Uber's systems allowed unauthorised (dismissed or suspended) drivers to upload their photographs to other Uber driver accounts, resulting in at least 14,000 uninsured trips across the city, according to TfL. They also did not have confidence that similar issues may not happen again, and said the company is "not fit and proper at this time".

There will be an appeal, but do you spot the real problem here? It's the drivers, best to get rid of them really.

So in 2016 an announcement was made, Volvo Cars and Uber were to develop autonomous driving cars. Both companies used the same base vehicle, the XC90, for the next stage of their own autonomous car strategies. Rather than having that stupid Google 'Firefly' car, Uber suddenly have a proper one to work with. Here was the future. A vehicle you may never have to drive, robotically turning up at your front door operated by a company that technically has never owned or run a car.

For the consumer a Volvo might be a nice way to travel, especially as there is no reason to actually own one.

Sounds great, but the technology has to work properly. On 18th March 2018, an autonomous car operated by Uber during real-world testing with an emergency driver behind the wheel, killed Elaine Herzberg probably the first recorded fatality involving a self-driving vehicle. Herzberg stepped into the road with her bicycle.

The National Transportation Safety Board in the States said that Uber's software failed to recognize the 49-year-old victim, as a pedestrian crossing the street, despite spotting her six seconds before the impact. It didn't calculate that it could potentially collide with her until 1.2 seconds before impact, at which point it was too late to brake. Uber's system design "did not include a consideration for jaywalking pedestrians." On top of that, the car initiated a one second braking delay so that the vehicle could calculate an alternative path or let the safety driver take control.

Uber's test vehicles may have failed to identify roadway hazards in at least two other cases, according to their report. In the seven months prior to the fatal crash, Uber vehicles were involved in 37 accidents, including 33 in which other vehicles struck the Uber test cars.

Uber also partnered with Ford and fellow app based taxi company Lyft. The stated aim is the creation of "SharedStreets, a new data platform designed to make it easier for the private sector to work with cities around the world and leverage data to improve urban mobility. The data sets pledged by the companies will provide the public and private sectors with new tools to manage curb space in order to reduce congestion and emissions that cause climate change; improve the efficiency of city streets by making it easier for everyone to get around; and save lives by preventing traffic crashes."

Let's hope this prevents any crashes, but it demonstrates how the traditional car making industry and the new fangled tech companies are getting together to lobby local authorities and national governments to bend to their will. On the face of it this is all about sharing statistics, drop offs and pick ups, but this has more to do with taxi monopolies.

Then there is Waymo, which has also equipped itself with some decent vehicles. Fiat Chrysler Automobiles joined up with Waymo in 2017, providing the self-driving startup with Pacifica minivans. Then they ramped up the cool car agenda by linking up with Jaguar to include their electric I-Pace.

Waymo committed itself to launching a driverless-taxi service in Phoenix Arizona, where it has been testing its cars. Unfortunately only part of that city is covered and only approved users can take part. Many believe that the wide, clear, bright streets amount to probably the easiest place on earth to drive any car. Not only that, Waymo's cars quite rightly have safety drivers behind the wheel. Not very straightforward all this self-driving business is it? In late 2019 Morgan Stanley Bank, cut its valuation of Waymo by 40%, to $105bn.

Closer to home, headlines in 2018 proclaimed that there would be self-driving cabs in London by 2021 said the BBC and Sky News, to name just a few. A little bit of digging reveals that these may not be Johnny Cabs, more like Bohnny Busses.

Before Uber, there was Addison Lee, still a huge player in the London cab market, but best known for offering motorcycle couriers and people carrier minicabs. They linked up with a University of Oxford tech company offshoot Oxbotica to develop driverless technology for their vehicles. It isn't specifically for their minicabs, but for bigger shuttle type buses. Likely to be on a fairly short and straightforward route. The thing is, London itself is far too fiddly and full of pedestrians and cyclists, so full self-driving cab machines are a very long way off.

Then again, you can just borrow or share a car. This is the cool way to hire a car. Combine a fancier name than Hertz with a hipster approach to carpooling, caring and sharing, plus an app of course and here we have the thoroughly modern way not to own a car.

The most high profile example of this is Zipcar which came about in 2000 with rentable cars parked around London (and more than 10,000 nationally). They say half the city's population are within a five-minute walk of their nearest zippy vehicle. So sign up for the app, locate the nearest Zippy, which is used to unlock the vehicle. It is then hired a car by the hour, or the day and returnable to any designated bay. They operate all over the world and there are dozens of similar schemes.

Also piling into the whole car sharing business are manufacturers who are obviously concerned where their future income is going to come from. Ford, already in bed with Uber and Lyft, went and bought Chariot. It used Transit Connect vans and sought to pool passengers partly by acting like a bus. But smaller, though still picking up passengers at the official stops. Effectively this would be the outsourcing of the bus service to Ford, oh and Uber and Lyft.

Meanwhile Ford's Go Drive initiative in London was shut down in 2016 because they don't see a future in car sharing vehicles that might be inactive some of the time and would otherwise be used to clog the streets. BMW and Mercedes would beg to differ.

Introduced in 2019 Share Now offered car sharing via an app called Drive Now, which means Londoners can share a ride in a Mini or BMW in Westminster, Tower Hamlets, Southwark, Barnet, Brent, Hackney, Haringey, Islington and Waltham Forest. This brought together Drive Now and Car2go under a single roof, with 4m users already registered with the service

Then of course you could join all these huge multinational companies and possibly beat them at their own game by sharing your car. Sounds a bit mad, but there are inevitably loads of apps you can sign up with like hiyacar.co.uk which are peer to peer schemes. Why you would ever want to trust someone else with your car, even if there is insurance and other so-called safeguards in place, is beyond a true enthusiast.

Then again, it might be possible to make a case for filling up some empty seats in your vehicle if you felt like spreading the cost. That's called carpooling. There is still element of trust and I'm not sure it is always appropriate at all, unless you know the people involved or work with them. Also tell your insurance company.

Which brings us right back to Johnny Cabs. That is the corporate dream, but it seems to be very hard to get there. The first stage is getting motorists out of their cars. Uber and similar services are doing that because on the face of it, they are so convenient and relatively cheap. The economics of owning a car in a major city is being undermined by parking and environmental regulations, leaving conventional public transport and the Johnny Cabs.

Total Recall was set in 2084 AD and so far we haven't colonised Mars, or got a cab to complete a journey on its own. Maybe it will never happen. Let's hope so.

*C02 in trees can be deadly...*

## 16. Carbon Neutral - *CO2: It's a naturally occurring gas that gives life. Discuss.*

I think we have already established what comes out of a car exhaust apart from the nitrogen, water and oxygen is pretty crappy. The thing is that CO2, which also pops out, does not seem to have a very good reputation and it is very hard to work out why it has been comprehensively demonised. Maybe the people who don't like carbon don't know very much about basic science. That's probably because those people are politicians or unelected busy bodies with an environmental and political agenda.

Let's keep all this to cars as much as possible because the link between cars and harmful carbon emissions has been repeated to the point when it can no longer be questioned. It should be. The point of this chapter is to establish that the production of carbon is largely irrelevant and inevitable result of life on earth.

Carbon Dioxide is literally the food of life and a very good thing because plants love the stuff. Specialist fruit and vegetable farmers actually pump CO2 into their greenhouses to promote growth of their crops. Plants build their tissues by using photosynthesis to take carbon from the air around them, it is that simple.

We are told that CO2 controls the climate. However, at 400ppmv (0.04%), it is a minor warming gas (water vapour is the predominant one).

Human activity are only the originators of maximum 4% of annual global $CO_2$ emissions; the other 96% of its emissions being natural and actually beyond our control and highly variable.

Meanwhile, the overwhelming determinants of terrestrial climate: solar activity and its interaction with the Earth's cosmic ray (low-level cloud-forming) environment, plus numerous other extraterrestrial factors, are in a large part ignored.

$CO_2$ is a climate tracker, not a climate determinant. When chronologically-synchronised, historical averaged global temperatures and carbon dioxide levels are tracked over billions of years, temperature changes always precede corresponding changes in atmospheric carbon dioxide levels by between 200 and 800 years (due to the massive heat capacity of the Earth's oceans; which warm - and cool - only very slowly).

Back to where we started. The anti $CO_2$ propaganda has got so bad that it is labelled as a pollutant, rather than being regarded as an essential trace gas for life on Earth. Without it, this world would be a lifeless mud-ball. Below about 170ppmv of carbon dioxide, plant life is no longer sustainable; leading to the consequent rapid extinction of all animal life (including humanity). The whole planet is a giant $CO_2$ recycling factory and long may that continue.

If vehicles and general manufacture is a carbon problem, what about humans themselves? If you applied the same rules to people as vehicles, then all of us would be banned from those City centres adopting low emission rules. OK, we might not be banned, but certainly be required to pay a daily charge to walk around an Ultra Low Emission Zone, but we shouldn't give them ideas. The average human produces around 900 grammes of CO2 per day according to the United States Department for Agriculture. Others statistics suggest that human pollution amounts to at least two tons of carbon dioxide a year each from the time food is produced to when the human body excretes it, representing more than 20 percent of total yearly emissions. That's according to a study by the Spanish Foundation for Science and Technology in 2010.

Even if we didn't go for the biggest figure the very average family of two adults and two children would produce 3.6kg of CO2 per day which adds up over a year to 1314kg from merely being alive and breathing. That may sound like a lot, but compared to the CO2 emissions of cars of 120g/km currently this equates to around 6850 miles annually, pretty much the national average mileage.

All the statistics surrounding CO2 are complicated and dependent on so many factors including, distance travelled, number of seats and which way the wind may be blowing at the time. It is far easier to just agree that CO2 is a good thing and leave it there. For a bit of fun though, let's have a look at comparing apples and pears. Actually the Energy Saving Trust did this for us in 2018 when they looked at a journey from London to Edinburgh by a slightly sad single person using various transport options, so we will run with that where possible.

**CO2 vs. Planes** This is the sexy one, a fast, efficient jet aircraft, it would nice if it was a private one, but let's stick with a great big scheduled one. Apparently Airplanes are about 3 percent of total global climate emissions. A single flight produces three tons of carbon dioxide per passenger, but that amount goes up if the plane isn't choc full. Planes also produce vapour trails and emit tropospheric ozone, which have an impact on climate, which isn't as long lasting as the CO2. Anyway the flight distance is 336 miles it would take one hour 25 minutes, but that doesn't include the any of the queuing, transfers and general security based mucking about you have these days in airport hell holes. The total emissions would be 144 kg

**CO2 vs. Trains** This is a very interesting one as it is going to depend on whether the train is diesel or electric, quite where the electricity comes from and just how many passengers are squeezed on board. According to the European Environment Agency, rail travel accounts for 14 grams of CO2 emissions per passenger mile. No account is made of building the train, its age and the amount of maintenance required. Anyway, the London to Edinburgh distance is 392 miles and the emissions would be a paltry 29kg.

**CO2 vs. Buses** On the whole any form of public or group

transportation is always going to be less emissions heavy. In this case it wouldn't be a double decker, but some sort of fairly fancy and comfy coach. American environmentalists would regard this as the ultimate in carpooling, plus you can take a nap, read a book and scoff sandwiches. Trouble is The Energy Saving Trust must have had something against buses or coaches, as they didn't cover that. I did though come across a company called Kings Ferry who used a carbon footprint calculator (National Atmospheric Emission Inventory) to prove that you would be doing the world a favour. So a trip from London to Edinburgh on one of their coaches results in average CO2 emissions per passenger of 9.2 kg per journey, whereas the same journey by a car carrying 1.56 persons would produce 71 kg, by plane 96.4 and by modern high speed electric train 11.9. These don't match up with the other figures, which sort of proves the point that there are so many variables.

**CO2 vs. Real Cars** The simple fact is that cars actually become a more efficient mode of transport with more people in the vehicle. So if you are planning on going to Edinburgh, take some friends or family, it's a lovely place. Obviously the emissions rate per person decreases. All of sudden a car turns into a train when it comes to emissions. So that's 29 kg CO2 x 4 - so 116 kg CO2, which is on a par with the petrol (120 kg CO2) or diesel car (115 kg CO2) emissions. It is enough to make you want to squeeze another one on board, even if you didn't like them much. Wouldn't be as comfy. Just to recap, that would be a 405-mile road trip and with a diesel car it would be 115kg and a petrol 120kg with just the one person going to Edinburgh. There is no allowance for the age of the vehicle and how efficient or otherwise the engines are. Also it is easy to see how some might see the diesel as rather more friendly to the environment compared to the petrol. That's how the whole diesel misunderstanding happened, but hey, there is more misunderstanding to come. Hybrid wasn't included in those stats but we did a quick calculation and with something like a Prius would be something like 106g/kg.

**CO2 vs. Electric Cars** We are poking people with electric sticks, or Tasers here and calling conventionally engined cars real ones. Electric cars are real, just powered in a different way. Technically it has zero exhaust emissions, as there isn't an exhaust. The power has to come from somewhere, so that would need to be factored in. Also long distance travel isn't an EV strongpoint. The majority of models on sale today would potentially take a couple of days. The original figures did allow for a four-hour charging session, which meant the total journey would take eleven hours. Recharging speeds depend on the type of charger used, but it could mean an overnight stop. The total carbon emissions of an electric car will vary depending on the source of electricity. However you do any calculations it will have a smaller carbon footprint than the real cars. Plus if you charge from a renewable energy source, such as solar or wind, in theory then it will be very low. The thing is though, nothing is zero and it is so tempting to factor in the incredible amount of carbon used to make a windmill or solar farm, never mind the EV itself. Let's call the electric car to Edinburgh, not zero g/km. I think that's fair.

**C02 vs. Boat** The Energy Saving Trust didn't cover boats, presumably because it wouldn't be a very direct route and might well take a day or so. Boats, container ships and cruise ships get a very easy ride when it comes to emissions. Whereas the motorist gets fined and banned, these floaty things are ignored. Amazing as they have such big, dirty, oil burning engines. Ships are responsible for roughly 3% of global CO2 emitting approximately 1 billion tonnes of CO2 and GHGs (greenhouse gases) per year. Indeed, it has been estimated that just one container ship, can produce the same amount of pollution as 50 million cars. There was a study of 203 cruise ships sailing in Europe in 2017, carried out by the European think-tank, Transport & Environment. They calculated that sulphur dioxide emissions from cars was 3.2m kt versus 62m kt from cruise ships and at the time there were 260m cars in Europe. Boats and ships are terrible aren't they?

**Carbon Offset** Having chosen your preferred mode of transport, but not a ship, then if you really cared about all this and wanted to participate in what seems to be the most colossally huge scam, you'd get yourself one of these very clever carbon offsets.

On the surface carbon offsets involve putting aside some money for renewable energy development to balance out the emissions you are generating. Either that or planting some trees to eat up some carbon. You simply go to a website tap in your journey co-ordinates and it comes up with your carbon footprint based on how much you drive and fly, and then it sells you offsets.

Historically, the U.N.'s clean development mechanism was the first major offsetting scheme, set up under the 1997 Kyoto Protocol, in which 190 countries agreed country-by-country emission reduction targets. There are approved projects all around the world and it re-directs money their way. Mind you there is no guarantee that the trees, which have been planted, will grow at all.

As a result there is such a thing as carbon trading. There was a Climate Change Roundtable formed in January 2005 at the Word Economic Forum. In that group were global companies like British Airways and British Petroleum, no surprise there, but Toyota and Ford were also among their number. If you wondered why there is so little, or just zero push back against the whole carbon as pollutant issue, by the motor industry there's your answer. Car companies are part of the whole carbon credit movement.

The stated aim was to act on climate change by offering what they called 'market based solutions'. They wanted governments to work towards 'clear, transparent and consistent price signals' through the 'creation of a long-term policy framework.' Bit of a word salad there, but essentially these multinational companies are not on our side, they want motorists and consumers to pay more money to governments: because 'climate'. They specifically mentioned greenhouse gases, which is a buzz phrase, which doesn't get used anymore.

A carbon credit is a permit or certificate allowing the holder, usually a company, or a member of the royal family or Hollywood celebrity to emit carbon dioxide or other greenhouse gases. The credit limits the emission to a mass equal to one ton of carbon dioxide.

Car companies are up to their corporate necks in all this, especially in America. The trading of carbon emissions credits has become a very big business. In 2012 the Environmental Protection Agency and the National Highway Traffic Safety Administration mapped out a greenhouse gas emissions reduction plan, so that each year manufacturers had to get cleaner.

General Motors in particular their Chevrolet cars was the leading corporate purchaser of carbon offsets in 2012 and 2013, buying credits representing the reduction of 4.6 million metric tons of greenhouse gas emissions. We have covered this elsewhere but the buying and selling of credits amongst themselves is little more than a money merry go round which presumably benefits bankers profits as much as it does the car companies. Back in 2014 Daimler AG's Mercedes-Benz arm acquired about 1.4 million credits, including 500,000 additional credits from Nissan.

More recently in America, Ford Motor Company linked up with TerraPass to market branded carbon offsets to all drivers of Ford, Lincoln, and Mercury cars and trucks. Apparently, this is Ford created carbon and the money goes into selected renewable energy projects sold under their Greener Miles program. Ford says they get no revenue from the sale and is funding the program as part of their broader initiative on climate change.

At the moment the Financial Services Authority in the UK does not regulate any climate trading activity and consumers should be wary as there seems to be a lot of scams aimed at ordinary members of the public. Potential investors are contacted by email, post, or word of mouth. They are offered carbon credit certificates, voluntary emission reductions (VERs), certified emission reductions (CERs) or an opportunity to invest directly in a 'green' scheme or project that generates carbon credits as a return on investment. Unless you are a virtue signaling celebrity trying to feel good about using a private jet, chances are, it is a scam.

In the real world these schemes don't seem to work at all and seem very scam like anyway. One prominent example is FIFA who for the 2014 football World Cup in Brazil, bought carbon credits covering 331,000 tonnes of carbon dioxide. There was a bunch of righteous projects in Brazil. One was a reforestation in the state of Rondônia, which had to be stopped when loggers destroyed more trees than all the credits sold.

Carbon is all about cash. That cash is not for you, but

extracted from your pocket and transferred into the coffers of Government and the world's biggest companies.

Volcanoes deliver far more carbon dioxide that man-made greenhouse gases and are going off all the time (For instance, Mount Etna adds about 16,000 tons of CO2 to the atmosphere each day, or 5.8 million tons per year).

Governments should tax volcanoes.

And very finally, prove that CO2 is bad and win cash money. The New Zealand Climate Science Coalition has offered the Professor Augie Auer prize of $10,000, to be granted to the first applicant to present real-world evidence showing that the man-made fraction of airborne carbon dioxide causes dangerous global warming. Full details can be found here: climatescience.org.nz

*Cycling Environmentalists are a genuine road hazard...*

## 17. Climate Deranged Syndrome - *How and why they plan to keep the red flag flying.*

Here's the problem, every single car company and motoring organisation in the United Kingdom and beyond has completely signed up to the climate change agenda. Maybe they are correct, it would certainly be impossible to sell any cars if they were not on board with what every government in the world believed. It does not matter what their customers might actually want, be it more affordable, practical and stylish vehicles.

Here is what the global car companies think about climate change and their place in it. Yes, it is very worthy, boring and environmental buzzword bingo kicks in pretty quickly, but stick with it.

FCA (Fiat Chrysler Automobiles) believes: *"That effective, lasting solutions to climate change and other pressing environmental issues can only be achieved through an integrated approach that combines individual and collective commitment; an effective multi-stakeholder strategy; and investment in enabling processes and technologies."*

BMW said: *"We are systematically reducing our environmental footprint in the area of energy usage, by continually enhancing the energy efficiency of our processes. We also evaluate the use of wind and hydropower at our sites, for example, and purchase renewable energy. Our long-term goal is to obtain 100%*

*of our electricity from renewable sources."*

Over at Daimler Benz: *"Through our initiative "The Road to Emission-free Driving" and the reduction targets it sets for our fleet emissions, we are helping to protect the planet from the effects of climate change."*

Connie Hedegaard, Member of the Volkswagen Sustainability Council said: *"By appointing the Council, Volkswagen has sent a clear signal: Climate change is a key issue for us. We consider this as a great chance to create an open dialogue space between Europe's biggest car manufacturer, scientists, environmental activists, and NGO's. Ever since, we've had some genuinely frank exchanges with the management board."*

According to Toyota's Testimony before the U.S. House subcommittee on Energy and air Quality: *"Toyota recognizes that climate change is occurring and ... is committed to continued action to address climate change and promote greater energy diversity..."*

And finally: *"Honda recognizes climate change as a serious environmental concern for all of society. For years, the company has been committed to mitigating climate change throughout our corporate activities."*

Trawling through car manufacturers official documents and statements to pick just a few comments was very depressing and repetitive. Car manufacturers are too scared to mount a counter argument, or suggest that pollution could be tackled differently. They feel obliged to go along with agenda because all the governments are as one on this issue. So how did we actually get here?

This might be the point at which you skip forward a chapter. Luckily we are near the end of Demotorization and you might have got the gist of what I'm on about. Bullying motorists to get more money and control is in every government's DNA.

Quite simply, control energy and you control civilization. The big conspiracy theory would be, any lowering of the world's CO2 levels causes crop yields to go down, and maybe people die off from starvation. That makes populations easier to control for the self-appointed elite of our planet. What they have very cleverly achieved is creating something so terrifying that everyone in the gullible world can rally against.

Creating the concept of 'global warming' suggested that if everyone did not work together then they would die, of warmth. Not only that, anyone who does not agree with this world view and therefore want to destroy the planet would also be guilty of being anti-science and just plain stupid.

However, the global warming narrative didn't work mainly because the planet did not seem to be warming at all. That explains why it was rebranded as 'climate change'. That too has become a fairly amorphous term after a few years so has been ramped up to become a 'climate emergency'. That's where we are right now. Any disaster, weather related or not, be it a hurricane or wildfire, drought or flood, are all signs that the world will end soonish. That's a decade to 12 years in the future at the time of when Demotorization came out.

In the meantime 'they' want to completely reorganise the way that society and in particular the economy works. Oh and change the way we drive. The important question is who 'they' actually are. I think we might well find them within the United Nations

Let's give full credit to the late Canadian, Maurice Strong. He was a former banker and oil tycoon and also the Secretary General of the 1972 United Nations Conference on the Human Environment. That event launched the whole environmental activist movement and his 1992 Rio Environmental Summit set the modern climate agenda. Plus he was the first Executive Director of the United Nations Environment Programme (UNEP). Whilst working for them he helped found the Intergovernmental Panel on Climate Change. The IPCC, which still effectively tells governments what to think when it comes to the climate. That is also the organisation, which is constantly quoted by those who reckon that the end of the world is near.

Rewinding back to Rio in 1992, as Conference Secretary-General, Strong said, "Isn't the only hope for this planet that the industrialized civilization collapse? Isn't it our responsibility to bring that about?"

Another significant speech at Rio came from Christiana Figueres, Executive Secretary of United Nations Framework Convention on Climate Change: "This is the first time in the history of mankind that we are setting ourselves the task of intentionally, within a defined period of time to change the economic development model that has been reigning for at least 150 years, since the industrial revolution. That will not happen overnight and it will not happen at a single conference on climate change, be it COP 15, 21, 40 - you choose the number. It just does not occur like that. It is a process, because of the depth of the transformation."

They, or shall we name them as the UN and their globalist helpers, that's multinational companies and most governments, have been and continue to play the long game. How do we know this? Maybe there is a document, which might help.

**The United Nations Agenda 2030** is the roadmap to the end of our prosperous western lifestyle, as we know it and scrapping of the motorcar. According to the UN themselves, "This Agenda is a plan of action for people, planet and prosperity. It also seeks to strengthen universal peace in larger freedom." Its full title actually is 'Transforming Our World: the 2030 Agenda for Sustainable Development' and it came into effect in January 2016. The 'Declaration' agreed upon at the United Nations meeting in New York has 53 points. Rather than boring you to tears with it all, because it is repetitive and does indeed promise everyone in the world a complimentary golden fleece, plus skies full of rainbows and biodegradable fairy dust, let's extract the points that are going to be pertinent to the motorist. The rest is too terrifying to contemplate.

**Ensure access to affordable, reliable, sustainable and modern energy for all.** What the UN want to do is demonise and also penalise the reliable and let's face it, 100% organic, fuel sources that have served us so well for generations, coal, gas and oil. Fracking and we won't go into any detail at all, is only controversial because the media says it should be. Meanwhile subsidies for 'green' energy solutions are readily funded by governments and that means tax payers. These are schemes that otherwise would not survive in a real world. For example, one of President Obama's biggest green failures (there were quite a few) was Solyndra LLC, a solar-panel maker. Presidential aides pressured White House budget officials to complete a review of a $535 million U.S. loan guarantee to the firm, which subsequently filed for bankruptcy protection in 2016.

In Britain there has been some disquiet from traditional energy suppliers like British Gas. Going back to 2015 when green taxes were set to cost households almost £150 and British Gas blamed the Government for a huge rise in electricity for three million of its customers. At the time Michael Gove, the Environment Secretary, suggested that more wind farms may need to be built to power a new generation of electric cars under government plans to ban the sale of diesel and petrol vehicles. Owen Paterson, Tory MP and former Environment Secretary, said: "It is the most regressive form of taxation since the Sherriff of Nottingham, transferring money from those who are the least well off to wealthy landowners and businesses. It's robbing the poor to pay the rich."

The Office for Budget responsibility, the fiscal watchdog, has forecast that environmental levies will rise from £4.6billion in 2015-16 to £13.5billion by 2022. Those are all old figures now so the only way is up for subsidies under the energy agenda.

**Make cities and human settlements inclusive, safe, resilient and sustainable** The UN would much prefer it if everyone lived in densely packed, tightly controlled cities where we would be under twenty four hour surveillance. Easy to boss about and making sure that no one had any personal means of transportation would trap them in a ULEZ hell hole. That seems to be what they are creating right now. Doing the groundwork and making no exceptions for anyone who dares to live in rural areas. Areas that have no transport infrastructure whatsoever. Darren Shirley, Chief Executive of Campaign for Better Transport, said in 2018: "Rural public transport is in crisis. Bus cuts and shrinking transport networks are making it harder and harder for people to get to work or school, to visit friends and family, or access shops and services, as well as putting extra pressure on our congested roads."

**Ensure sustainable consumption and production**

**patterns** This is more about taxing fossil fuels and electricity then making people put up with increasingly worse standards of living. Shaming people because they dare to use their unfair share of petrol or electricity. According to The Organisation for Economic Co-operation and Development in 2018 the United Kingdom performs well in most measures of well-being relative to most other countries in the Better Life Index. The United Kingdom ranks above the average in personal security, environmental quality, civic engagement, social connections, health status, jobs and earnings, income and wealth, education and skills, and subjective well being. It is below average in work-life balance and housing. There is then, room for improvement.

**Take urgent action to combat climate change and its impacts** Less cars, that's what this means. It is easy. Or at least getting everyone to change their cars for something that is electric, because they tell us that those will save the planet. The alternative is mass public transport, not perfect, but necessary to fight that nasty climate change.

According to the Committee on Climate Change, which advises the government on building a low-carbon economy and preparing for climate change "The 2008 Climate Change Act made the UK the first country to establish a long-term legally binding framework to cut carbon emissions. It contains a target requiring emissions reductions by at least 100% by 2050. To limit the most damaging impacts of climate change, we need to reduce greenhouse gas emissions globally."

Here then is a very real agenda to get you out of your car and into a 1984 style UN administered urban hellhole. For our well being of course. However, what are they actually protecting us from? The real problem is that the climate doom mongers are not very good at making predictions. If we listened to them, well we would have all given up by now. It is simply a case of following the fashionable buzzwords, some of which are not so fashionable, or buzzy anymore.

**The inconvenient truths...**The '60s and '70s were all about the coming Ice Age. Earth Day was first celebrated on April 22nd 1970 and was a suitably depressing event. If you haven't heard of it, this is an American invention. It was like Christmas, but without the food, jollity, Santa Claus, or presents. It has been an excuse for environmentalists to share their predictions and much is made of just how wide of the mark everyone has been since then. Briefly, they were fairly sure that the end of civilization would happen in 15-30 years. As a build up to that, hundreds of millions of deaths from starvation in the coming decade, were all topped off with a new ice age by 2000.

They liked Ice back then and one of the most prominent articles was in Time Magazine in June 1974. It focused on African droughts, record rain in Pakistan and Japan, plus poor crops and chilly weather in Canada. Most remarkably of all they said rainy Britain was drier than usual. All of this with increasing pack ice around the island of Iceland all added up to a new Ice Age, which so far, hasn't happened. It could still occur and things climatically are quite chilly at the moment, but Time thought it was imminent back then.

Well, that didn't frighten anyone, but that hole in the ozone layer sounded plausible. Especially if you shaved, or teased your hair, which was pretty much the entire population. Bottles of hairspray and cans of shaving cream contained chlorofluorocarbons (CFCs) which a 1974, Nobel Prize winning paper showed that the CFCs used in spray bottles destroyed atmospheric ozone. That's the invisible layer (that's always important, like CO2 you can't see it) surrounding the Earth. That meant there was no protection from UV rays so that we might go blind, or at the very least least get sunburn. No need to tick the manual, or electric sunroof option back then when buying a car. Concern about this stretched in the '80s, what with all those complicated 'Flock of Seagull' hairstyles. So CFCs were banned. NASA says the hole has been patched up, despite us being told was unfixable. Scientists now admit it's a seasonal phenomenon that forms during Antarctica's spring, when weather heats up and reactions between CFCs and ozone increases. There are apparently little holes elsewhere, like above Tibet. Panic over.

Global warming then became the new scare and it was associated strongly with ice caps melting and sea levels rising. The Maldives is supposed to be an underwaterworld by now, but you can still go there on holiday. In June 1989 the United Nations Environment Program Director Noel Brown said that entire nations could be wiped off the face of the earth by rising sea levels if the global warming trends are not reversed by the year 2000. He said that governments have a 10-year window of opportunity to solve the green house effect before it gets beyond human control.

This is all sounds so familiar. Brown said if the warming trend continues, "the question is will we be able to reverse the process in time? We say that within the next ten years, given the present loads that the atmosphere has to bear, we have an opportunity to start the stabilizing process."

At some point Global Warming wasn't working as a catchphrase and Climate Change was adopted, around 1994. As a phrase It fitted in with these sorts of stories from the Independent in March 2000 when they ran with a story that 'Snowfalls are now just a thing of the past.' Charles Onians reported that Dr. David Viner, a senior research scientist at the Climate Research Unit based at the University of East Anglia, when he said that snow would be 'a very rare and exciting event ...Children just aren't going to know what snow is.'

Then in February 2004 The Guardian reported that the Pentagon told President Bush that 'Climate Change will destroy us'. They also added that Britain will be 'Siberian' in less than 20 years and specifically printed a date: 2020. Just for fun they also added that Nuclear conflict, mega-droughts, famine and widespread rioting, will erupt across the world. Well, someone is always rioting about something, otherwise when I last looked through the car window, Norfolk wasn't quite Siberia.

Back to The Independent in July 2009 when they broke their 'don't do royals' self-imposed embargo to indulge the opinions of Prince Charles. The heir to the throne told an audience of industrialists and environmentalists at St James's Palace that there was 96 months, presumably because it sounded less than eight years, to save the world. The head of the Duchy of Cornwall Waitrose Organic biscuit brand, blamed capitalism and consumerism. Delivering the annual Richard Dimbleby lecture, Charles said that without "coherent financial incentives and disincentives" we have just 96 months to avert "irretrievable climate and ecosystem collapse, and all that goes with it."

Equally deluded and self important was Prime Minister Gordon Brown, also speaking in 2009 when in October he addressed the Major Economies Forum in London and cut the end of the world prediction to a nice round 50 days. Or least the preparation for the postponement of the end of the world... "There are now fewer than 50 days to set the course of the next 50 years and more. So, as we convene here, we carry great responsibilities, and the world is watching. If we do not reach a deal at this time, let us be in no doubt: once the damage from unchecked emissions growth is done, no retrospective global agreement, in some future period, can undo that choice. By then it will be irretrievably too late."

2009 was a bumper year for making solid gold, copper bottomed, nailed on certainty future predictions. Also still around and also not a scientist, like Prince Charles or Gordon Brown, but a consummate politician is Al Gore. His previous wife Tipper made all the best '80s and '90s albums easy to spot in old fashioned record shops because of the parental advisory sticker, his talent was peddling numerous climate inaccuracies.

Gore was addressing the U.N. climate conference in Copenhagen in December when he suggested that the Polar Ice Cap may disappear by the summer of 2014, "It is hard to capture the astonishment that the experts in the science of ice felt when they saw this," said the former Vice President. One would assume that they are still astonished today to find that the cap is still on the North Pole.

Gore had form at doing this, indeed he made a film all about it. 'An Inconvenient Truth' remains one of the worst films to win an Oscar although that is no guarantee of quality. It was no 'Withnail and I' and it was supposed to be a documentary, but sadly didn't contain much in the way of actual truth, Indeed, the High Court ruled on it in England.

Lorry driver Stewart Dimmock brought legal action against the school system to prevent 'An Inconvenient Truth' from being taught in school as factual information.

The court adjudged it to be propaganda containing scientific inaccuracies and "sentimental mush" If teachers present the film without making this plain they may be in breach of section 406 of the Education Act 1996 and guilty of political indoctrination. Never mind, at least beachfront mansion owner Gore is probably pleased that the rising sea levels prediction has not come to pass. Lots of other things didn't come to pass, including, a new ice age, flooding in China and India, a parched South Sahara and no more Polar bears.

Why would you ever trust any organisation or government which says all it needs is most of your money to make the weather better. Firstly their record on all this is terrible. Then there is the assumption that they can change the weather, or indeed anything at all. Governments routinely promise the earth in their election manifestos and then a few years later fail to deliver.

There have been signs that some of us are not falling for the greenwash and there has been a backlash.

Trust the French to do it properly and get togged up in their High Visibility jackets to become the legendary Gilets Jaunes. Except that the world's media has largely ignored them. A proposed fuel tax rise started it all. That's because the French government is committed to reducing its CO2 emissions. That fuel-price hike would be used to finance renewable-energy projects and discourage the use of diesel and petrol cars. Incredibly, the price of diesel had already risen by 23 per cent in the space of a year. This hit the rural dwellers hard, as they rely heavily on their cars. They also hated the cut in the speed limit on rural roads from 90km/h to 80km/h. Many believed was a way to make money from speeding tickets. No wonder that the yellow vests targeted 60 per cent of the country's speed cameras, by taping them up, burning and knocking them over. This is the French working class versus the intellectual elite.

'The government talks about the end of the world. We are talking about the end of the month.'

There are no identifiable leaders of the Gilets Jaunes and they have been subject to the most horrendous police brutality in the face of massive provocation. Inside the movement is a broad coalition of political views, but they are united by their working class roots living outside of the municipal areas and appalled that the elites can take such high handed and selfish decisions.

Here is living proof of the UN agenda in action. The rural working class is marginalised whilst all the money is being pumped into the cities and metropolitan areas. 60% of the population is being ignored, but at least they are fighting back.

Other grassroots movements in Europe have been led by farmers. The Germans brought Berlin to a halt with their tractors in a claim that new environmental limits being planned are overly restrictive and that the government is making it impossible for domestic agriculture to compete against imports. The French farmers did not want to be left out so did the same in Paris about the banning of specialist pesticides; there were also protests in Belgium.

Ordinary people in the West are being told to change the way they live and what they drive. They are being asked to do this for the good of the planet and being taxed heavily in the process. Meanwhile in China they are building coal fired power stations by the week and the rate of consumption in India and South America rises year on year. Good for them. Westerners giving up cars and living miserably won't have any effect. It is merely a massive and pointless virtue signal that will change nothing.

So whatever they are calling it this week, be it Global Warming, Climate Change, or Climate Emergency, it is a colossal con. If you can go to your local bank and still get a 25-year mortgage to buy a house with a garage attached, then the world's not going to end, just yet.

*Government Vehicle Confiscation Service*

## 18. The Car. Discontinued. By force. - *They really are coming for your car.*

Maybe they know Demotorization the Book is coming, but the authorities seem to be accelerating towards a private transport free future.

**The Politicians** The first, "Let's ban cars claxon" was loudly sounded by The Science and Technology Committee which highlighted the lack of Government policies in place to deliver the net zero carbon emissions target by 2050 and recommends 10 steps the Government should take to meet this legally binding target. It is grim little document and it is all summed up by Rt Hon Norman Lamb MP, Chair of the Committee who said: "Parliament has declared a climate emergency. The worrying effects of climate change, such as heat waves, wildfires and flooding are already occurring at an alarming rate and will have a huge impact on future generations."

Let's look at their Plan for reducing vehicle emissions: "The Government must bring forward the date of its proposed ban on the sales of new 'conventional' cars and vans to 2035 at the latest, and ensure that it covers hybrids too. In the near-term, the Government must reconsider the fiscal incentives for consumers to purchase both new and used vehicle models with lower emissions. The Government should also work with public services and owners of public land, such as schools and hospitals, to accelerate the deployment of electric vehicle charge points, and introduce measures to ensure that charge points are interoperable, compatible with a smart energy system, reliable, and provide real-time information on their current functionality. Although ultra-low emissions vehicles generate very little emissions during use, their manufacture generates substantial emissions. In the long-term, widespread personal vehicle ownership therefore does not appear to be compatible with significant decarbonisation. The Government should not aim to achieve emissions reductions simply by replacing existing vehicles with lower-emissions versions."

Several things here. They are terribly excited about the coming ban on what they call conventional cars and also have a problem with hybrids. They may not be perfect, but it is a technology, which has made electricity not just acceptable but practical. Never mind, because the magic electric cars will save the day, what we need are loads more charging points in publically owned areas. Finally there seems to be an acknowledgement that even electric cars create all sorts of emissions in production, so technically were never zero emission. That however, presents us with a real problem when they say private vehicle ownership in the future is 'incompatible'.

That was the opinion of a cross party group of MPs, but as that's a future promise, made by politicians, it may not happen. The thing is that there is now an environmental pushback against electric cars. Simply being ‘lecky' is no longer decent protection against the green zealots. However squeaky clean your vehicle it will never, be clean enough.

The political tide has been turning against the motorist and the car for some time. As early as November 1973, the Marquess of Hertford argued in the House of Lords that, “It is painfully obvious that petrol will increase in price so that, even if there is no shortage or rationing, the idea of driving for pleasure will become as extravagant as it is eccentric.” In January Labour MP Frank Judd proposed in the House of Commons that “In view of the on-going and long-term energy crisis, which is likely to be with us for all time, would not the Minister agree that the time has come to call a halt to superfluous road building and to use the public funds thus saved to finance effective and efficient public transport.”

In February 1974, the chairman of the Transport 2000 study group (They were everywhere for a couple of decades, but exist now as The Campaign for Better Transport), Tony Blackburn (not the ancient DJ), argued that, “if the energy crisis means anything it means a complete change in transport priorities, with less emphasis on private motoring.”

Never mind the exhaust pipe, or lack of pipe, the new environmental concern is now dust from brakes and tyres. Apparently fragments of micro plastics from tyres, road surfaces and brakes will not just get into the air and you're your throat, it could also flow into rivers, and the sea, according to the government's Air Quality Expert Group.

There is a warning for us there as they said that "No legislation is currently in place specifically to limit or reduce [these] particles." Presumably there will be legislation and they also stated the blindingly obvious, "So while legislation has driven down emissions of particles from exhausts, the non-exhaust proportion of road traffic emissions has increased."

In theory then, all cars, even electric ones are for the chop. The proof of it happening right now can be found out there in the real world. First of all Bristol decided to make a name for itself by becoming the first UK city to ban diesel cars from its streets. The city council voted, following a public consultation, to impose a total ban on all privately owned diesel vehicles during the day in the city centre. This scheme, still needs government approval, is due to start in 2021. So there is time for this to unhappen.

Meanwhile in Birmingham The city's network of road tunnels could be shut down to cars and more steps taken to ban motorists as part of a 'clean air' master plan being drawn up by Birmingham City Council "Residential neighbourhoods and local centres will be places where people are put first. Cars will no longer dominate street life around homes and schools. Active travel – walking and cycling – will become how most people get around their locality most of the time.

"Parking will be used as a means to manage demand for travel by car through availability, pricing and restrictions. Where redevelopment potential exists, land currently occupied by car parking will be put to more productive use."

As has been pointed out in the ULEZ Chapter there are Cities queuing up to do this and there are new low emission recruits in Europe every day.

Then there are bicycles, everyone likes bikes. They don't take up much space, they get you fit, there are only human emissions. Wonderful. Except that bit where they don't take up much space. That seems to have changed rather a lot.

Traffic management is never easy and there is a point of view that sharing the same road space is a recipe for trouble. Separating drivers from bikeists has been around for a while. Go back to 1934 and Minister of Transport Lesilie Hoare-Belisha. He pops up occasionally in transport history, opened a section of the Western Avenue A40 between Hanger Lane and Greenford Road. You might have thought cyclists would have been pleased, but they feared being driven off the road completely and solely onto these designated sections of separated roads. By 1938 forty-five miles of new arterial roads had bike paths running beside them, but then the Second World War happened.

Building new towns in the post war era did make some accommodation for cyclists such as Stevenage and

Harlow, which got bike paths. In the north the Tyne Pedestrian and Cyclist Tunnel, was opened in 1951, sixteen years before the motor vehicle tunnel and still in use today.

Bicycles went out of fashion as a mass transportation in the '60s and '70s. It came back into fashion, obviously in London. In the '80s there was the London Cycle network backed by the Greater London Council. It was a collection of routes, via back streets between places rather than main roads using signposts and white lines on the roads. Cyclists thought it was rubbish and inadequate and slow.

Things moved on in 2010 to the creation of cycle superhighways. First of all it was blue paint on the ground. That moved on in 2016 to segregated Central London routes, effectively building two roads where there used to be one. The new cycle highways now take up a third more of the road compared to the previous bicycle lanes. This has increased congestion, traffic and pollution as motorists sit there idling. Cyclists may claim that the number of cyclists using these routes has increased by 60%, also more cyclists means less pollution and cyclists have never been safer.

What we do seem to have in London is something approaching gridlock. The Embankment has been the most high profile example of all day and night traffic jams. In 2017 according to Transport for London, almost 7,500 cycle trips were made along the Embankment each day before the protected cycle way was installed.

Since the creation of the cycle way the number of bike journeys increased to 10,300 a day.

I don't want to make this Londoncentric, it's a terrible place for cars and that's the point really. The slow death of a capital city by terrible traffic management. Installing cycle lanes almost randomly without thinking of the consequences is par for the course. It is probably right that two sorts of disparate vehicles don't mix. If you banned all vehicles, not just cars, how will goods be delivered to shops and homes? Not by scooters, but the bigger stuff to supermarkets and shops? By all means get rid of the car and prioritise the bicycle, but there needs to be some sort of proper planning. Otherwise the inevitable happens which is this...

Motoring data company Inrix was reported in the Independent Newspaper in 2019 that the average speed of traffic in London is an average of 7mph. That meant that London drivers waste 227 hours a year stuck in traffic. It is the most congested City and sixth globally. Birmingham, Glasgow, Manchester and Bristol, just so you know to avoid them are the other most congested cities in Britain.

**A Warning from Ireland** A country that is very nearby to Britain, also has plans to ban the sale of new petrol and diesel vehicles by 2030. That's ahead of the UK target at 2040 and the aim is to ensure that all new cars and vans on Irish roads in 11 years' time are electric vehicles. In order In order to avert the so called "climate apocalypse", the Irish government plans to force people "out of private cars because they are the biggest offenders for emissions", according to transport minister Shane Ross whose proposals that includes banning fossil fuel vehicles from towns and cities nationwide.

This proposed legislation was among 180 measures in their Climate Action Plan, so should we be concerned? Well, Ireland is a tiny country of less than 5 million, yet these draconian measures cannot be undone once implemented and it is only a matter of time before similar policies make the hop over the Irish Sea. For instance, the government said it would stop granting National Car Test (NCT) certificates to fossil fuel vehicles by 2045. Still a long way off, but that is the equivalent of making your petrol or diesel illegal. You won't even get a choice to keep on using your 15-year-old car, which is surely the environmentally sane thing to do.

**Education.** They are coming for your children. If you were sceptical over the assertion in Demotorized that the United Nations are up to their necks in all this climate change nonsense, well here's the proof. Britain in 2019 got its first United Nations-accredited climate change teacher. This has happened in North Devon, where Kingsley School's head of geography, Steve Whaley, is bringing completely unbiased analysis to young spongy minds. This isn't scary at all. The programme is free for all primary and secondary school teachers. "Once accredited they are able to deliver climate change facts and the best practices for mitigation — preparing young people to protect the planet."

**Media.** Sadly you can't trust any outlet on radio, TV, printed or online for a fair and balanced view, there always seems to be an agenda. Climate change has been accepted without question. Most worrying is that they do not present an alternative or balancing view. Nothing. The only mainstream publications, which consistently present a contrary view to the narrative is The Spectator and Spiked Online. What you can rely on newspapers and television for is to show pictures of cars and exhausts on a regular basis to illustrate pollution.

**Motoring media** It is worth singling out car magazines and programmes as a specialist case. They should be sticking up for the car and mostly they do. Some journalists believe the 'scientific consensus' and are onboard with climate change. Cars must adapt or die. Others completely disagree. At least there are some divergent views and room for debate. The thing is, no journalist matters. Especially in the shrinking print media where their influence is now marginal. Only one person truly counts when it comes to cars and that's the Patron Saint of car journalists, Jeremy Clarkson.

In 2019 he seems to have decided that climate change is something to believe in. When shooting Grand Tour he found that when paddling a boat on the Mekong River in Vietnam that one lake had shrunk dramatically.

"It's the first time that we've ever admitted to there being global warming," Clarkson told The Sunday Times. "It was alarming, genuinely alarming."

How he can make that assertion without knowing an awful lot about the geography and history of the region means that the statement is meaningless. He has fallen into the trap of saying that something, anything is due to climate change. It is interesting that he has used the old, outdated term, Global Warming. Maybe that's because at his age it is too late to be reprogrammed. What we need to remember is that Jeremy is part of the establishment. TV Royalty, a legendary columnist and author. Potentially there is a career to lose by saying the wrong thing. That has not stopped him in the past, but as I have already pointed out there is an international agenda, being pushed by the great and good. Not just politicians, it now includes celebrities like Jeremy Clarkson.

He comments about global warming are just that, his his opinion and he is perfectly at liberty to express them. How he intends to modify his behavior and lifestyle as a result of this revelation will be fascinating to watch. Presumably that means no more Grand Tour, or any television appearances at all.

It is impossible to find anyone presenting a television programme or starring in Hollywood films that has ever offered an opposing view on climate change. The best I can do is David Bellamy, who was as big as it gets when it comes to presenting science and in particular botanically based programmes and documentaries on TV. He was a scientist, environmentalist had a distinctive delivery and florid manner plus the biggest, bushiest beard on telly. He also had the audacity in 2004 to call global warming, 'poppycock'.

He was interviewed by The Independent newspaper in January 2013 just before he turned 80 years old. They were checking in with the highest profile climate denier to see if he had changed his mind.

"All of the work dried up after that. I was due to start another series with the BBC but that didn't go anywhere, and the other side didn't want to know. I was shunned. They didn't want to hear the other side."

Full credit to Bellamy, because he wasn't backing down, and still believed that global warming was bollocks, "Absolutely. It is not happening at all, but if you get the idea that people's children will die because of CO2 they fall for it."

Not really motoring media at all there with Bellamy, but I think the point is made. (Very sadly, he passed away after I had written this, but he led a remarkable life and was at the forefront of many environmental campaigns and initiatives).

Car magazines seem to like electric cars a lot, but you need to remember that they get lent cars for free, so they would wouldn't they? You also need to follow the money, because with plug in cars comes rather large advertising revenues. I don't think they are biased, but they like the idea of new technology and see it as the future.

I would cite the 6th of November 2019 Editorial by Mark Tisshaw in Autocar as proof that there is still genuine independence and insight in the nation's car magazines. The title is "EU CO2 Targets are killing City Cars. Here's why it must change". Yep, small, frugal, light and practical cars replaced by two tonne electric SUVs.

There is hope, even if they insist on taking our cars.

*Self Propelled Traffic Association*
*on manoeuvres...*

## 19. 'Self-Propelled Traffic Association' - *Don't let them Raise the Red Flag Again*

We return to the very beginning. That bloke with the red flag is back. He's not just slowing us down, he's ganged up with his political and environmental mates and is pulling us out of the driving seat. I believe this is what Demotorized proves.

If you object to being told what to do, how to drive then you have to tell someone about it. Ideally someone with a bit of clout. Like a politician. Except that they don't seem to be on the motorists' side. The Alliance of British Drivers certainly are, but never seem to get a fair crack of media time. I am half tempted to restart the Self-Propelled Traffic Association in honour of Sir David Solomons original 1895 campaign against the so-called Red Flag Act. However, that does seem a bit too much like hard work.

There is always the High Vis option. Regular weekend protests with like-minded individuals in town and city centres. It may involve the authorities using tear gas, batons and casual violence, but we don't live in France.

Alternatively, it is possible to use legal and illegal ways to show your displeasure. Possibly buy a tractor on the basis that they are permitted to use rebated heavy fuel oils or gas oil, the latter known as "red diesel".

You could borrow your Nan's mobility scooter. I always thought scooters had two wheels, but hey, these don't have to pay road tax on condition that they only reach a maximum speed of 8mph on the roads and are fitted with a device limiting them to 4mph on footpaths.

Steam power is another consideration because that is not liable for Car Tax. Maybe we could build our own electric vehicle, which might at least be affordable. Or just adopt an old milk float. So you never know, if Demotorized gets some traction, there may be a follow up with diagrams and practical advice on how to build a better self-propelled vehicle that won't be taxed to oblivion. That would be Remotorized.

Thank you for reading this book and please keep on Motorizing.

Very finally there are some questions and answers...

*The Sunny Uplands approach to the future of motoring...*

## 20. ...Widespread personal vehicle ownership... does not appear to be compatible... - *Always Question the narrative.*

In the spirit of peering into the future and to see if there actually is one for driving and cars, I asked five simple questions. These were directed at major motoring and transport related organisations, plus political parties. Not everyone replied. Some replied twice. A few organisations I had to remind. Several just ignored my polite requests. The political parties were very disappointing. Initially there was an election on, but after that they didn't have any excuse not to get back. This was supposed to be the second from last chapter, but so I could paste in any late replies it bookends, the book. All of these replies are posted without comment. Make up your own mind about how the future for motorists is shaping up.

**Roger Lawson Campaign Director The Alliance of British Drivers** This is a voluntary organisation promoting the interests and concerns of Britain's drivers. www.abd.org.uk

**1. What is your organisation's relationship with the British Motorist, if any?**

We try to represent the interests of the British motorist. There is a need to represent their interests in a world, which is becoming increasing hostile environment for those that choose to drive.

**2. How does your organisation view the British Motorist?**

They should do more to protect their own interests because they are currently overtaxed, bearing an excessive amount of the tax burden through the road fund licence and lots of indirect taxes.

**3. Are there any viable alternatives to these forms of private transport, which you support, or encourage?**

The alternatives to a private motor car are not always appropriate especially if you are disabled. Some, like cycling can also be dangerous. Public transport may be too slow, or just inconvenient.

**4. Is there anything you believe that the Motorist can do to improve their lot?**

Join our organisation, support the campaigns and get these issues in front of the elected politicians.

**5. Will there be room for the Motorist in the future?**

The stated aim of the government is to decarbonise the economy and that represents a threat to all forms of private transport. Effectively they would just want us to travel on electric trains. There seems to be little real evidence of climate change, yet the government seems determined to make us live in a medieval society. There is a future for motorists so long as they are prepared not to give in to government pressure and use their cars because they are convenient, practical and cost effective.

**Brian Gregory founder member and current Policy Director Alliance of British Drivers**

**1. What is your organisation's relationship with the British Motorist, if any?**

We aim to communicate the rational, scientifically literate road user's perspective on road transport issues; including road safety, environment, taxation etc.

**2. How does your organisation view the British**

**Motorist?**

We view the British Motorist and Motorcyclist as an enormous group of individuals that is unjustifiably subject to a steadily increasing tide of hostile, social-engineering-motivated, legislative, fiscal and environmental measures. This onslaught is borne out of ignorance, dogma and junk-science; and has no genuine scientific or moral justification. It is a politically-legitimised campaign of victimisation that currently generates enormous revenues - and holds colossal further, future revenue-raising opportunities created by the ever-tightening legislative ratchet to which road users are subjected.

**3. Are there any viable alternatives to these forms of private transport, which you support, or encourage?**

For the vast majority of non-intra city journeys , there are no realistic, cost-, time-, and energy-efficient alternatives to personal mobility; i.e., cars or PTWs. Bear in mind that time is THE only really irreplaceable asset of which any human being has (albeit temporary) possession.

**4. Is there anything you believe that the Motorist can do to improve their lot?**

Continue working to overturn the junk-science, soundbite culture that pervades the modern media; and seek to give the public properly-researched information about the pros and cons of largely unrestricted personal mobility.

**5. Will there be room for the Motorist in the future?**

If we do not fight for the survival of this right - hard fought for since Magna Carta times, and subsequently won - of unfettered access to mobility for everyone, we stand to lose it forever: because that is our opponents' end-game. So, if we lose this fight, NO. Everyone will be slaves to the Climate Change Nazis.

**The British Parking Association**

**1.What is your organisation's relationship with the British Motorist/Motorcyclist, if any?**

The British Parking Association represents the UK parking and traffic management profession. Our 750 plus corporate members includes local government, commercial providers and parking system operators, as well as consultants and academics. In addition, we represent over 500 individuals and have a separate membership structure to support their professional development.

We are a not-for-profit organisation working with our members and stakeholders to support their communities, improve compliance by those managing and using parking facilities, and encourage fairness to achieve our vision of excellence in parking for all.

**2.How does your organisation view the British Motorist?**

We view motorists and motorcyclists (and cyclists) as customers of parking services.

**3. Are there any viable alternatives to these forms of private transport, which you support, or encourage?**

We promote all modes of transport whether that be by bus, park and ride, bicycle, motorbike, train or tram. All road transport requires a parking space; bus depots, parking at out of town park and ride sites, bicycle and motorcycle parking and parking at railway stations. We envisage transport hubs that will support different types of transport to deliver a connected journey.

**4. Is there anything you believe that the Motorist can do to improve their lot?**

With more people traveling than ever before, there is a greater demand for road space and parking. Motorists have a responsibility to ensure they consider how using their vehicles impacts on the environment and to ensure they always drive and park in an appropriate way in accordance with the rules.

**5. Will there be room for the Motorist/Motorcyclist in the future?**

There will always be room for the motorist and motorcyclist. Cars and motorcycles may be switching from fossil fuels to electric power, but the requirement of individuals to be mobile will not diminish. The number of vehicles on our roads increases every year and so does the demand for parking. Whether more people switch to cycling, walking or public transport will be determined by investment and economic opportunities in terms of employment and housing.

**Transport for London. Christina Calderato is the Head of Transport Strategy and Planning**

**1.What is your organisation's relationship with the British Motorist, if any?**

We manage the busiest roads in London, known as the red routes, which make up 5 per cent of the capital's roads but carry 30 per cent of traffic and spend over £400m a year running this network. Keeping everybody safe on our roads, including people driving cars and riding motorbikes, is our number one priority.

Our 24/7 Control Centre works closely with the emergency services, local councils and other partners to make minute-by-minute decisions about our roads, tunnels and traffic signals, helping to keep everybody in London moving, including people driving and riding motorbikes.

**2.How does your organisation view the British Motorist?**

We value all of our road users and don't categorise them into groups – in fact, many people who drive also walk, cycle and use public transport.

We make decisions based on safety and other

important issues for Londoners, including congestion, air quality, the environment and health.

Our number one priority is keeping everybody safe, whether they are driving a car, walking, riding a bike or catching the bus. That's why we're working towards our Vision Zero goal of eliminating all deaths and serious injuries on our transport network, including people driving cars and riding motorbikes.

**3. Are there any viable alternatives to these forms of private transport, which you support, or encourage? Will there be room for the Motorist in the future?**

As London continues to grow, demand for our limited road space will continue to grow, which is why we need to make sure that it is used as efficiently as possible. We also need to make sure that the vehicles travelling on our roads don't make our air more polluted.

Walking, cycling and public transport take up much less space on the roads than cars, which is why the Mayor has set out a target of having 80 per cent of journeys made by walking, cycling and public transport by 2041. London's public transport is affordable and easy to use, we are improving the environment for people walking, and our safe, high-quality Cycleway network continues to grow across the capital.

Enabling more people to use these ways of getting around leaves more space on the roads for people who do need to drive. Our world-leading Ultra Low Emission

Zone has also cut nitrogen oxide pollution by a third in central London, reducing the impact of car traffic on Londoners' health.

**4. Is there anything you believe that the Motorist/Motorcyclist can do to improve their lot?**

We want everybody – including people cycling, driving and walking – to look out for one another when using our roads, so that we can end the tragic impact that collisions on our roads cause.

People riding motorbikes are particularly vulnerable on our roads, which is why we're working hard on a range of measures to keep them safe, including offering a range of motorcycling training courses and bringing in the world's first Direct Vision Standard for lorries so that they can be seen more easily.

**5. Will there be room for the Motorist/Motorcyclist in the future? NOT ANSWERED**

**Motorcycle Industry Association. Nick Broomhall Communications Manager.**

**1.What is your organisation's relationship with the British Motorcyclist, if any?**

We are the industry body representing all Powered Light Vehicles (PLVs), also known as L-Category vehicles. The largest L-Categories are L1 (mopeds) and L3) motorcycles.

**2.How does your organisation view the British Motorcycist?**

Users of powered two wheelers (PTWs) are diverging into two distinct groups.

Traditional motorcycles are often (although not exclusively) leisure vehicles, with their riders being older enthusiasts, who would describe themselves as motorcyclists or bikers and to whom bikes are an important part of their life.

In parallel, commuters and light cargo operators are using increasing numbers of PTWs and other PLVs, a trend that is likely to increase as single-occupancy private cars and lightly laden vans are excluded from urban settings.

**3. Are there any viable alternatives to these forms of private transport, which you support, or encourage?**

MCIA is watching developments in the micromobility field and feels that regulated, legal-to-use e-scooters etc. have a place in the right situation.

MCIA is an advocate of "The Right Vehicle for the Right Journey"

**4. Is there anything you believe that the Motorist/Motorcylist can do to improve their lot?**

Motorists could ditch their cars for urban journeys and use a space efficient, low/zero emission PTW or PLV, with cars reserved for extra urban and long journeys.

**5. Will there be room for the Motorist/Motorcyclist in the future?**

It's hard to see a place for private cars (especially single occupancy) in urban settings of the future, as even an electrically-powered line of 5m x 3m boxes is still a traffic jam. Better use of that road space would be a reallocation towards light vehicles and the removal of safety issues caused by shared road space, PTWs and PLVs have a big role to play in (especially) urban travel.

**Demotorized References**

0. Hansard. DVLA, BBC, The Sun 1. On Four Wheels Orbis Publications • Autocar • Seven Sisters Anthony Sampson • OPEC • Forbes 2. Esso • Shell • HMRC Gov • Phys.org 3. Hansard • Sucking Eggs Patricia Nichol 2009 • Peter Edwardson speed limit.org • Live Science • Wall Street Journal • House of Commons Report Taxation of Road Fuels • Fuelling Oil Demand Transport and Environment Department • ITV News 4. On Four Wheels Orbis Publications • Hansard5. On Four Wheels Orbis Publications • Twickenham Times • Home Office • Transport for London • Alliance of British Drivers • Derbyshire Constabulary • Politics.co.uk • The Daily Telegraph • History of Road Safety gov.uk 6. Association of British Parking • National Car Parks • Highway Code • Daily Telegraph • The Guardian • BBC News • RAC Foundation • The Sun7. Kent Live News • RAC Foundation • Essex Live • Hansard 8. Transport For London • Auto Express • New York Times • Talk Radio.com • The Guardian • Mark McArthur Christie mmcmusings.com • Theconversation.com 9. Vance Packard • Unsafe at Any Speed Ralph Nader 1965 • Driving Standards Agency • Auto Editors Consumer Guide • The Encyclopedia of Classic Cars Thunder Bay Press 10. Vehicle Standards Agency • Vance 'The Wastemakers' 1960 • Society of Motor Manufacturers and Traders • Transport for London 11. New Scientist • Dyson.co.uk • The

Independent Newspaper • C. Simons, "The Lies of Unleaded Petrol", Nexus, April/May, June/July 1995. • P. Sawyer, Green Hoax Effect, Groupacumen Publications, Victoria, 1990. • Peter Sawyer, Green Hoax Effect, Groupacumen Publishing, Wodonga, Victoria, Australia, 1990 • New Scientist July 1986 • Pick myself Up and Dust Myself off by Colin Caudell 12. On Four Wheels Orbis Publications • Autocar • UK Gov Vehicle Licensing Stats • New York Times • BBC.co.uk 13. On Four Wheels Orbis Publications • BBC.co.uk • Uber.com • Waymo.com • U.S. National Highway Traffic Safety Administration • Thatcham Research • Association of British Insurers • Huffington Post • Forbes • New York Times • Wired UK • The Economist • The Independent • The I Paper 14. On Four Wheels Orbis Publications • Hansard • Battery vehicle Society • Who Killed The Electric Car 2006 • Toyota.co.uk • Autocar • Clean Technica • Dyson.co.uk • SEAT.com • Autocar • The Spectator • Green Car Reports • The Hill • U.S. Department of Energy 15. Evening Standard • Daily Mail • Mail Online • Public Carriage Office • Transport for London • National Transportation Safety Board • Ford • Waymo • Fiat Chrysler Automobiles • BBC • Sky News • Automotive News Europe • The Guardian 16. Transport for London • Energy Saving Trust • Environmental Protection Agency and the National Highway Traffic Safety Administration • climatescience.org.nz • European Parliament News • Carbon Brief • Autocar • Stopthesethings.com • Science Daily • The Guardian • Brian Gregor • RAC.co.uk 17. Fiat Chrysler Automobiles • BMW • Daimler Benz • Volkswagen • Toyota • Honda • United Nations • The Guardian • The Independent • Natural News • Climatism.com • Mail Online • principia-scientific.org • Competitive Enterprise Institute.org • Zerogedge.co 18.Birminghamlive • Parliament.uk • Electroverse.net

**Demotorized index** 

**Books by James Ruppert**

**World Cup Cortinas** In 1970 something very remarkable happened. Not only had England qualified for the World Cup in Mexico because they were defending champions, but most important of all Ford had loaned every squad member the greatest saloon car in the world. A book about what happened to those Ford Cortinas and how footballers bought better cars when they earned more money and no longer owned the same motors that your mum and dad drove.

"Channeling the absurdist spirit of Spike Milligan" said Classic Car, others just called it "Witty", that's **My Mini Cooper Its Part in My Breakdown** which has the full history of the world's favourite small car, plus the history of what came before but not after.

The Award Winning and so good the BBC based a documentary (Das Auto) on it **The German Car Industry, My Part in its Victory**. Winner of the Mercedes-Benz Montagu of Beaulieu Trophy for the best motoring book of year it contains jokes and references to champagne fuelled excess and 1980s music.

The Critically acclaimed and best selling **The British Car Industry Our Part in Its Downfall**, which has been described as "Beautifully written with considerable wit" by the Daily Telegraph and The Daily Express said, "It's informative, well-written and extremely funny" which was really nice of them.

The commonsense guide to motoring the **Bangernomics Bible**, which is future proof and universe friendly. Thatmeans it won't date ever even if you buy a hover bike or whatever is the latest thing is in 2059 and wherever you are in the world it just contains the common sense advice you need to buy a decent secondhand car. Classics Monthly said "Witty Book full of sound advice about buying a cheap safe and reliable runabout.

Here is a Diet programme for the car owner who is fed up piling on the pounds. Instead, The **Bangernomics Diet** promises to change the way we buy and run cars, for the better forever. Saving large amounts of Beans (money) by running a Banger (car). However, not just any old Banger, maybe a Classic or a Posh car bought for buttons.

Right back in the beginning, 1993 there was Bangernomics and now there is **Bangernomics Rebooted which** contrasts the absurd expense of buying a new car with the supreme good sense of buying well used. At a stroke depreciation no longer became an issue, running costs were slashed and there were no finance charges to be endured. Here is the original book, republished and effectively rebooted.

**Find these on www.bangernomics.com**

**James Ruppert** once sold cars. Lapsed member of the Battery Vehicle Society. Contributes every week to Autocar. Founded Free Car Mag and Bangernomics Mag. Drives Bangers.

Printed in Great Britain
by Amazon

35380077R00184